Plankton in a Changing World

Albert Calbet

Plankton in a Changing World

The Impact of Global Change on Marine Ecosystems

 Springer

Albert Calbet 🆔
Institute of Marine Sciences - CSIC
Barcelona, Spain

ISBN 978-3-031-76120-1   ISBN 978-3-031-76121-8   (eBook)
https://doi.org/10.1007/978-3-031-76121-8

This Springer imprint is published by the registered company Springer Nature Switzerland AG
The registered company address is: Gewerbestrasse 11, 6330 Cham, Switzerland

If disposing of this product, please recycle the paper.

About the Book

"Plankton in a Changing World: The Impact of Global Change on Marine Ecosystems" is a detailed and accessible examination of the crucial role plankton play in the world's ocean and the significant challenges they face due to global environmental changes. This book complements "The Wonders of Marine Plankton," which explored, in a casual and entertaining way, the intricate details of plankton ecology and their peculiarities and surprising characteristics. In "Plankton in a Changing World" the focus is on how global changes impact these essential organisms and, in turn, the broader marine ecosystems they sustain. While this book stands on its own, familiarity with "The Wonders of Marine Plankton" can enhance understanding of some chapters, making the complex interactions and impacts even clearer.

The book opens with an introduction to plankton, providing a comprehensive overview of the various types of plankton and their ecological importance. It covers everything from the foundational phytoplankton, which are key to oceanic primary production, to the complex interactions within planktonic communities. These initial chapters set the stage by explaining the diverse world of marine plankton and their critical functions in ocean ecosystems.

Following this, the book explores the distribution and habitats of plankton, emphasizing the variability in biomass across different ecosystems and the unique strategies plankton use for dispersal and survival. This section highlights the dynamic and often patchy nature of plankton distribution, shaped by both biological interactions and physical oceanographic processes.

At the heart of the book, we find an in-depth analysis of the impacts of global environmental change on marine plankton. With increasing ocean temperatures, acidification, and altered nutrient dynamics, plankton face numerous challenges. The book examines these issues, discussing the effects of global warming, the role of plankton in the ocean's biological pump, and the consequences of smaller plankton in warmer ocean.

Human activities and their effects on plankton are also addressed, including pollution, overfishing, and the introduction of classic and emerging pollutants. These chapters highlight the complex interplay between human-induced changes and plankton dynamics, emphasizing the need for sustainable practices to preserve these vital components of marine ecosystems.

The book concludes with case studies and regional perspectives, offering insights into plankton in various climatic zones, from polar regions to tropical waters. This section provides a global viewpoint on the diverse responses of plankton to environmental changes, illustrating the different challenges and adaptations in various regions.

Finally, the future directions section brings together the key themes of the book, discussing the importance of public awareness and education in promoting plankton conservation. It also outlines emerging trends and future research directions, emphasizing the need for advanced methodologies and technological innovations in studying and monitoring plankton.

"Plankton in a Changing World" is an essential read for anyone interested in marine life and environmental science. While it is designed to be accessible to the general public, the depth and coverage of information make it a valuable resource for marine biologists, ecologists, and environmental scientists seeking to understand the critical role of plankton in marine ecosystems and the profound impacts of global change on these tiny but mighty organisms.

Contents

About the Author

Albert Calbet is a marine researcher at the Institute of Marine Sciences, CSIC, in Barcelona, Spain, specialized in the ecology and ecophysiology of micro- and mesozooplankton. His work has significantly advanced our understanding of the role of microzooplankton in marine food webs. Albert earned his Ph.D. in Marine Sciences in 1993 from the Institute of Marine Sciences (ICM), CSIC, followed by postdoctoral research at the University of Hawaii at Manoa. At ICM, he has held various positions, including Deputy Director.

Albert has published over 130 peer-reviewed articles, authored several books and book chapters, and actively participated in scientific conferences worldwide. He has also been involved in teaching and mentoring students at the Ph.D., Master's, and undergraduate levels. His research has been supported by prestigious institutions, and he has served as a reviewer for funding agencies and on the editorial boards of high-impact scientific journals. Dedicated to science outreach, he manages several web pages and engages with the public through social media, outreach articles, conferences and books.

Part I

Introduction to Plankton Ecology and Major Groups

1

An Introduction to the World of Marine Plankton

The whole of the world's ecosystems are based on the healthy ocean and if that part of the planet becomes dysfunctional and goes wrong, then the whole of life on this planet will suffer
—Sir David Attenborough.

Marine plankton, those tiny organisms drifting in the ocean and seas, form the foundation of aquatic ecosystems. Despite their small size, they play a colossal role in maintaining the health and functionality of marine environments. Plankton are broadly categorized into groups such as virioplankton, bacterioplankton, microzooplankton, phytoplankton, mixoplankton, zooplankton, and ichthyoplankton. Each group is home to a multitude of species, each contributing uniquely to the marine ecosystem. In this chapter, I will present a brief summary of the major groups and functions of plankton, which will be further elaborated upon in the following chapters.

© The Author(s), under exclusive license to Springer Nature Switzerland AG 2024
A. Calbet, *Plankton in a Changing World*,
https://doi.org/10.1007/978-3-031-76121-8_1

Phytoplankton: The Ocean's Primary Producers

Phytoplankton are the primary producers in the marine food web, akin to terrestrial plants. Through photosynthesis, they convert sunlight into energy, producing oxygen and organic materials that sustain nearly all marine life. Despite their lower biomass compared to terrestrial plants (1 vs 450–470 Gt C, respectively), phytoplankton produce roughly the same amount of oxygen as terrestrial plants. However, most of this oxygen is readily consumed by marine organisms, and only a very small fraction, if any, reaches the atmosphere. Phytoplankton group includes various types of algae, with diatoms (Fig. 1.1) and dinoflagellates being among the most prominent. Phytoplankton are integral to the carbon cycle; they absorb carbon dioxide during photosynthesis, helping to mitigate the greenhouse effect and regulate the global climate. When they die, some of their carbon sinks to the ocean floor, effectively sequestering it and reducing atmospheric carbon dioxide levels. Additionally, phytoplankton serve as a primary food source for zooplankton, thus sustaining a complex and diverse marine food web. Their abundance and productivity also influence the distribution and health of fish populations, which are vital for global fisheries. Moreover, phytoplankton affect nutrient cycling by transforming inorganic nutrients into organic matter that other organisms can use, playing a pivotal role in maintaining the balance of marine ecosystems. Through their various functions, phytoplankton support biodiversity, contribute to biogeochemical cycles, and influence the health and stability of the entire oceanic environment.

Zooplankton: The Ocean's Primary Consumers

Zooplankton, the animal-like counterparts of phytoplankton, are critical as primary consumers in the oceanic food web. They feed on phytoplankton (and bacteria, smaller zooplankton, etc.) and, in turn, provide nourishment for larger marine animals, including fish, whales, and seabirds. Zooplankton encompass a diverse range of organisms, from microscopic protozoans to larger forms like jellyfish and krill. Protozoans include a varied array of groups, including ciliates, flagellates, dinoflagellates,

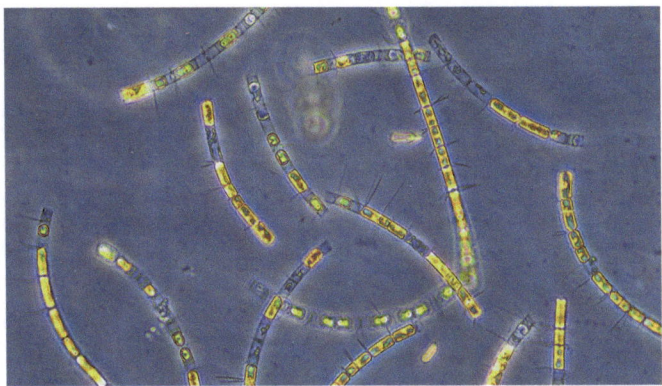

Fig. 1.1   Marine diatoms (*Chaetoceros pseudocurvisetus*). © Albert Calbet

Fig. 1.2   Acantharian (Radiolaria). © Albert Calbet

foraminifera, and radiolarians, among others. Some of them have naked bodies, while others possess protective shells. For instance, foraminifera and radiolarians (Fig. 1.2) have intricate calcium carbonate or silica shells that contribute to ocean sediment after they die. These organisms are key indicators of environmental changes and past climatic conditions.

Fig. 1.3 The marine copepod *Calanus minor*. © Albert Calbet

Protozoans and some small metazoans are known as microzooplankton and they play a key role as primary consumers, feeding on phytoplankton and bacterioplankton, transferring energy and organic matter up the food web. They are also vital in the microbial loop, consuming bacteria and recycling nutrients, thus maintaining balance in the ecosystem.

Among zooplankton, copepods (Fig. 1.3) and krill are particularly abundant and ecologically significant. Copepods, often referred to as the "insects of the sea," are crucial for many fish species and are the most abundant metazoans on Earth. Krill, small shrimp-like crustaceans, are vital in polar regions, forming dense swarms that serve as a primary food source for larger marine animals, including whales, penguins, fish, etc.

The Interplay Between Phytoplankton and Zooplankton

The interaction between phytoplankton and zooplankton is a delicate balance that supports the entire marine ecosystem. This dynamic relationship ensures the flow of energy and nutrients through the food web.

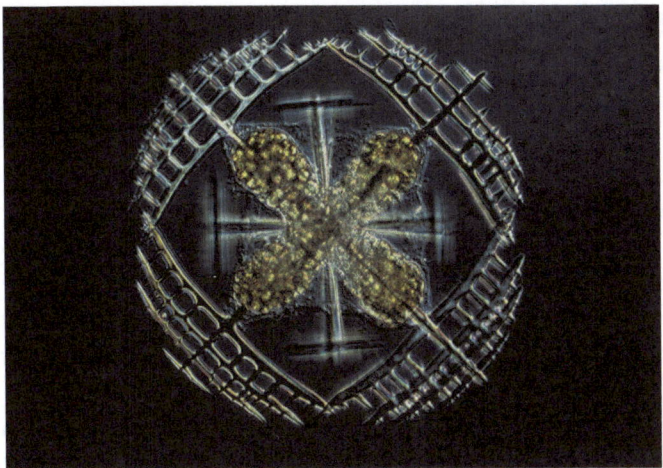

Fig. 1.4 The mixotrophic acantharian *Lithoptera* sp. © C. Carré

Phytoplankton absorb carbon dioxide during photosynthesis, acting as a significant carbon sink and playing a vital role in regulating the Earth's climate. When zooplankton consume phytoplankton, the carbon is transferred up the food web and eventually sequestered in the deep ocean as these organisms evacuate fecal pellets or die and sink to the seabed. This process, known as the biological pump (Chap. 18), is essential for mitigating the impacts of climate change.

Mixoplankton: The Dual Nature of Plankton

Mixoplankton (Fig. 1.4), alike the terrestrial carnivorous plants, are fascinating organisms that combine characteristics of both phytoplankton and zooplankton. They can photosynthesize like plants but also consume other plankton as animals do. This dual capability allows them to adapt to varying environmental conditions, making them important players in nutrient cycling and energy transfer in the ocean. Their flexible feeding strategies help maintain the stability and resilience of marine ecosystems.

Virioplankton and Bacterioplankton: The Microscopic Recyclers

Virioplankton, viruses that inhabit marine environments, are incredibly abundant and play a crucial role in controlling microbial populations and recycling nutrients. By lysing their host cells, they release organic matter back into the environment, making it available for other microorganisms. This process, known as the "viral shunt," helps regulate the microbial loop, an essential component of the ocean's nutrient cycle. Virioplankton also influence genetic diversity and evolution through horizontal gene transfer.

Bacterioplankton, prokaryotic plankton, are pivotal in decomposing and recycling organic matter. These microorganisms break down organic compounds from dead plants and animals, converting them into inorganic nutrients that can be reused by other organisms, particularly phytoplankton. Their metabolic activities help maintain the balance of nutrients in marine ecosystems, supporting the productivity and health of the ocean.

Ichthyoplankton: The Future of Fish Populations

Ichthyoplankton, the planktonic stages of fish, include eggs and larvae critical for replenishing fish populations. These early life stages are highly dependent on suitable planktonic food sources, such as some large phytoplankton and zooplankton, for growth and development. The survival and success of fish larvae are influenced by various factors, including water temperature, salinity, and the abundance of predators and prey. Understanding ichthyoplankton dynamics is crucial for fisheries management and the conservation of fish stocks.

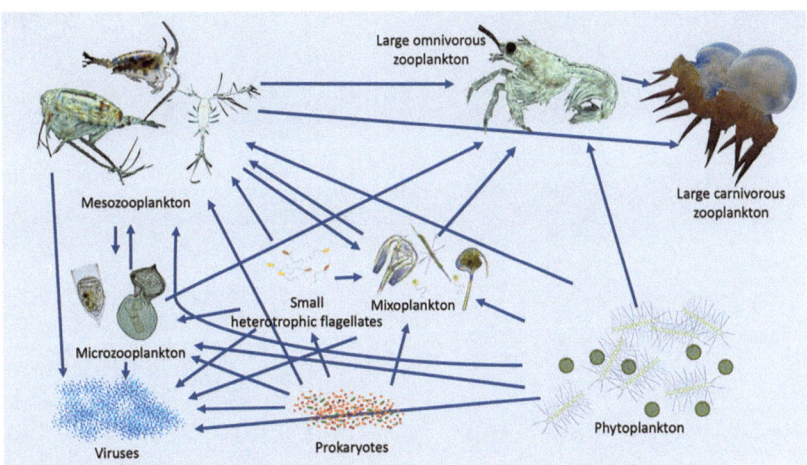

Fig. 1.5 Representation of the marine planktonic food web. © Albert Calbet

The Complex and Dynamic Marine Planktonic Food Web

The marine planktonic food web (Fig. 1.5) is a complex and dynamic system supporting a vast array of marine life. Phytoplankton produce organic matter through photosynthesis, which is consumed mostly by bacterioplankton, microzooplankton, and copepods. These, in turn, are preyed upon by larger zooplankton such as krill and jellyfish. Mixoplankton add complexity by functioning as both primary producers and consumers. Virioplankton control microbial populations that recycle nutrients, while ichthyoplankton rely on these interactions for their development and growth.

This food web is constantly influenced by environmental factors such as temperature, nutrient availability, and light conditions. Seasonal changes, climate variations, and human activities impact plankton abundance and distribution, affecting the entire food web.

Fig 1.5 Representation of the marine plankton food web

The Complex and Dynamic Marine Planktonic Food Web

2

Plankton Size Classification

Plankton come in all shapes and sizes, from microscopic creatures invisible to the naked eye to giants larger than a beach ball. But how do we categorize this incredible range? In the previous chapter we have seen a classification of plankton based mostly on their function and trophic modes. But scientists have developed also a system known as size classification to understand the ecological roles and challenges faced by these fascinating drifters.

A Historical Perspective

The concept of size classification for plankton emerged from early scientific explorations of the aquatic world. In the 1880s, German biologist Victor Hensen conducted pioneering studies on marine life. Using nets with varying mesh sizes (Fig. 2.1), Hensen captured different groups of plankton, laying the groundwork for today's size-based classification system (Fig. 2.2). Over time, scientists have refined this system, resulting in a spectrum that encompasses the entire range of planktonic life.

© The Author(s), under exclusive license to Springer Nature Switzerland AG 2024
A. Calbet, *Plankton in a Changing World*,
https://doi.org/10.1007/978-3-031-76121-8_2

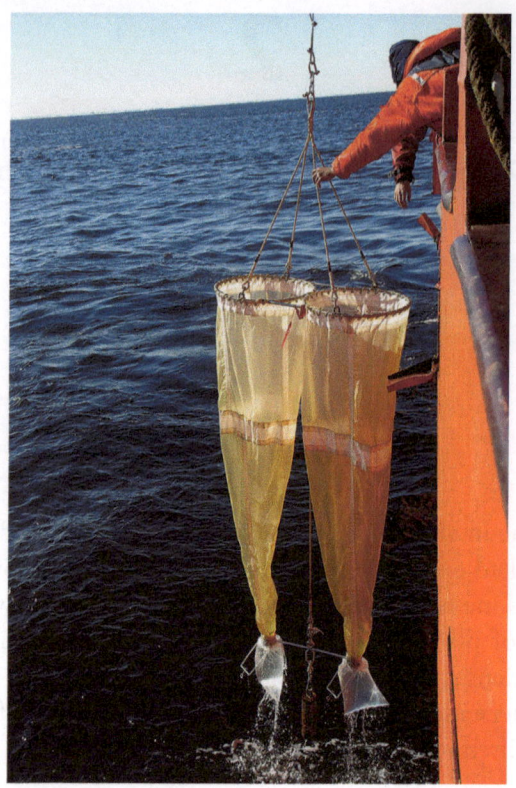

Fig. 2.1 Plankton net. This particular net is a double WP2, modified to hold a plastic bag instead of a cod end. This modification, combined with a very slow towing speed, allows for the collection of very fragile organisms. © Albert Calbet

The Size Categories

Megaplankton: The largest size in the classification, encompassing organisms exceeding 20 cm. These giants include jellyfish and some colonial animals that can stretch for meters, such as salps and siphonophores. Megaplankton are relatively rare and play specialized roles in the marine ecosystem.

Macroplankton: Ranging from 2 to 20 cm, this category features a wider variety of organisms, including krill and other large crustaceans.

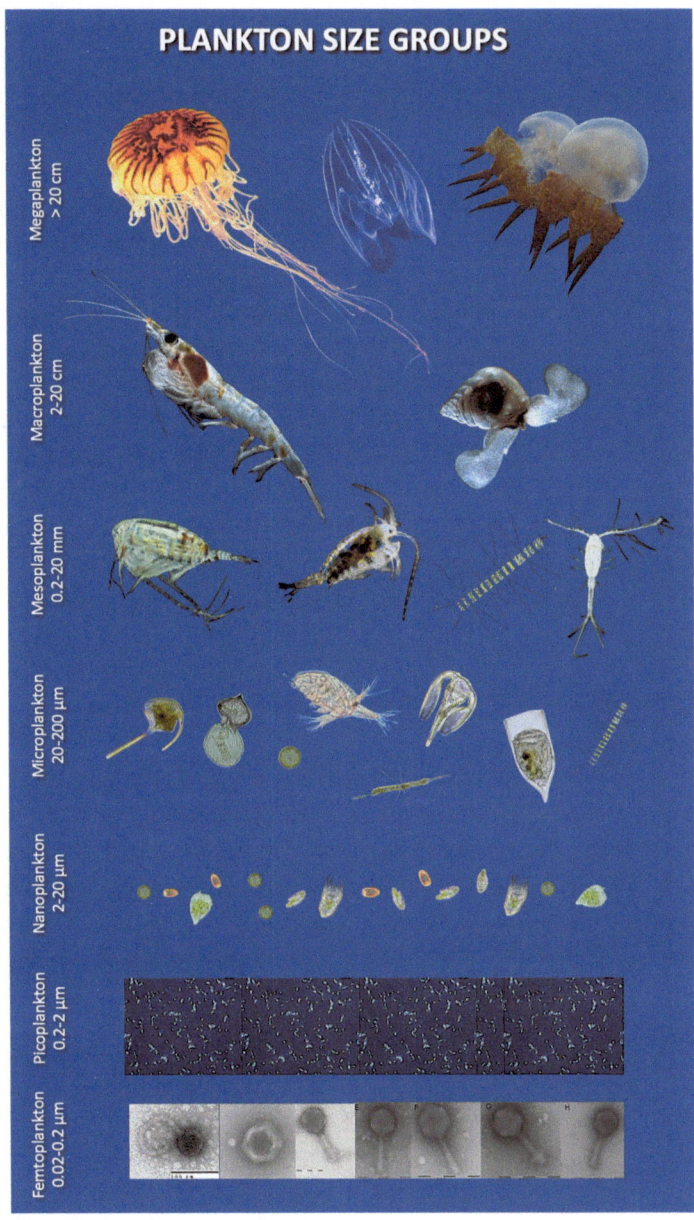

Fig. 2.2 The classification system for plankton based on size was first published by Sieburth et al. in 1978. The image, assembled from various sources and authors: A. Calbet, R. Hopcroft, D. Vaqué, E. Lara, Y.M. Castillo

Mesoplankton: Organisms between 0.2 and 20 mm represent a significant portion of the planktonic community. This size range, encompassing most copepods and some large phytoplankton, is crucial for studying energy transfer within the plankton ecosystem.

Microplankton: Sized from 20 to 200 μm, this category ventures into the microscopic realm. Dominated by phytoplankton and protozooplankton, microplankton play a vital role in primary production and carbon mobilization.

Nanoplankton: Even smaller, at 2 to 20 μm range, this size-fraction is dominated by flagellated protists. These microscopic organisms play a significant role in nutrient cycling and primary production.

Picoplankton: Pushing the boundaries of visibility, this category encompasses organisms between 0.2 and 2 μm, mostly bacteria and tiny flagellates. These minute entities are crucial players at the base of the marine food web.

Femtoplankton: The smallest category includes mostly viruses, which are key to controlling bacteria and phytoplankton populations.

Ecological Implications

The size classification system is not just about physical dimensions; it has profound ecological implications. Larger plankton tend to be solitary, while smaller plankton often exist in vast aggregations and high numbers. This difference influences how they interact with the environment and contribute to biogeochemical cycles. Additionally, the size of an organism dictates its susceptibility to predation and its physiological capabilities.

3

From the Bottom Up: Phytoplankton's Major Groups

In the first chapter, we discussed the critical roles of phytoplankton in marine ecosystems: they form the cornerstone of the marine food web, serving as primary producers by generating organic matter through photosynthesis. They are vital contributors to oxygen production and play a crucial role in the carbon cycle by absorbing carbon dioxide from the atmosphere. Additionally, phytoplankton are key components in nutrient cycles, such as nitrogen and phosphorus, transforming these elements into organic compounds that are readily consumed by grazers. Given their pivotal role, it is essential to highlight the major groups of microalgae that comprise this diverse and vital group.

Diatoms: The Silica Producers

Diatoms are algae characterized by their intricate silica shells called frustules, which offer protection (Fig. 3.1). Thriving in nutrient-rich waters, diatoms are a major component of phytoplankton blooms. Their beautifully patterned silica shells serve as protection against grazing, but also influence their buoyancy and light absorption capabilities. Diatoms are

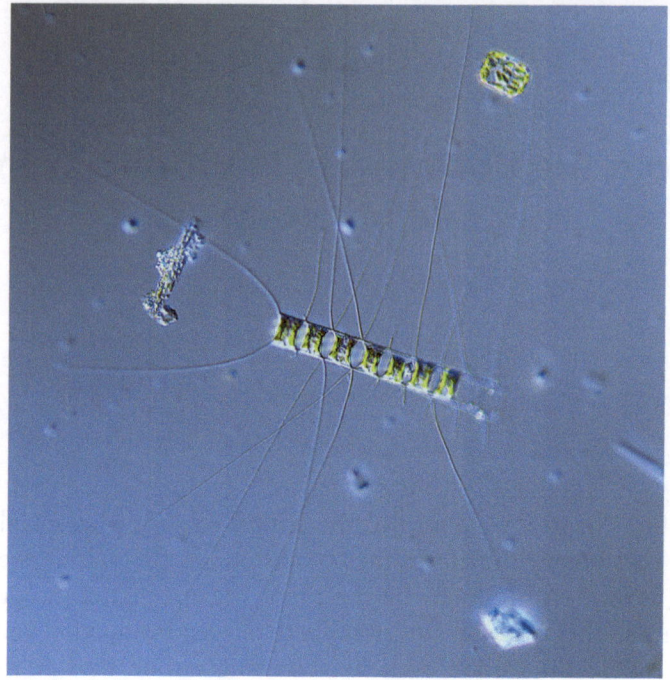

Fig. 3.1 Marine diatom. *Chaetoceros* cf. *decipiens*. © Albert Calbet

prolific and can be found in nearly every aquatic environment, from freshwater lakes to the open ocean. They play a crucial role in the ocean's silicon cycle due to their silica-based cell walls.

Dinoflagellates: The Dancers of the Sea

Dinoflagellates (Fig. 3.2), distinguished by their two flagella, are known for their diverse shapes and bioluminescence. They can cause harmful algal blooms, such as red tides, which produce toxins harmful to marine life and humans. Dinoflagellates exhibit a variety of forms and behaviors, with some species being photosynthetic and others heterotrophic, consuming other organisms for energy. The ability of some of them to

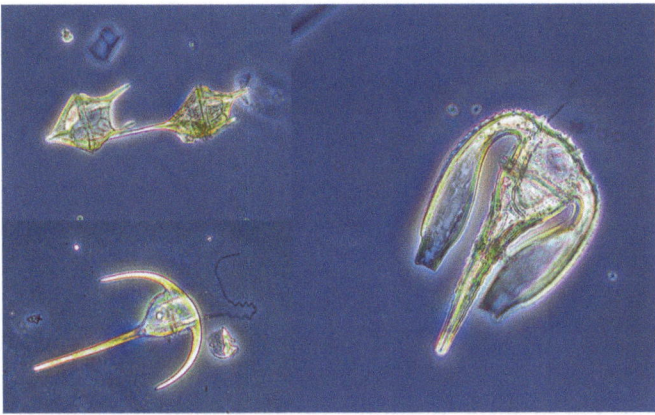

Fig. 3.2   Marine dinoflagellates. Left up: *Tripos pentagonum;* left down: *Tripos* sp., right: *Tripos platycornis.* Note that the genus *Tripos* was previously known as *Ceratium.* © Albert Calbet

produce light through bioluminescence is a fascinating adaptation, often seen in the mesmerizing glow of the ocean at night.

Coccolithophores: The Carbonate Architects

Coccolithophores, covered in calcium carbonate plates called coccoliths, contribute significantly to the marine carbon cycle. When they die, these plates sink to the seafloor, forming marine sediments and aiding long-term carbon sequestration. Coccolithophores are especially important in open ocean environments and are a key component of marine plankton communities. Their calcium carbonate plates not only protect the cell but also play a role in regulating the ocean's alkalinity.

Cyanobacteria: The Ancient Nitrogen Fixers

Cyanobacteria, also known as blue-green algae, are ancient organisms capable of nitrogen fixation, converting atmospheric nitrogen into a form usable by living organisms. They are among the oldest life forms on Earth and were crucial in shaping the planet's early atmosphere through oxygen production. Cyanobacteria thrive in a variety of environments, including freshwater, marine, and even terrestrial habitats, playing a vital role in nitrogen cycling, particularly in nutrient-poor waters.

Green Algae: The Coastal Powerhouses

Green algae (Chlorophyta) share similarities with terrestrial plants in their pigment composition and photosynthetic processes. While many species are found in freshwater environments, some also inhabit marine ecosystems. Green algae contribute significantly to the primary productivity of coastal waters and can form large blooms under favorable conditions. Their chlorophyll pigments give them a bright green color, making them an essential part of the food web by providing energy for various marine organisms.

Other Relevant Groups of Phytoplankton

Euglenoids are primarily found in freshwater but also in marine environments. They are motile and have a flexible cell membrane, allowing them to move toward light sources for photosynthesis. Some euglenoids can switch between autotrophic and heterotrophic modes of nutrition, depending on environmental conditions.

Prymnesiophytes (Haptophytes) often possess flagella and a unique structure called a haptonema. They are significant contributors to marine primary production and can form extensive blooms.

Cryptophytes are small, often overlooked algae that thrive in both freshwater and marine environments. Cryptophytes have unique pigments that allow them to photosynthesize efficiently in low-light conditions, such as those found in deep or turbid waters. They are an important food source for many zooplankton species and play a key role in nutrient cycling.

In summary, the diverse groups of phytoplankton play unique and vital roles in marine ecosystems. Understanding these groups and their functions is essential for appreciating the complexity and importance of marine food webs and biogeochemical cycles. The subsequent chapters will examine further these fascinating organisms and their ecological significance, particularly facing the threat of global change.

4

Ocean's Hidden Hardeners: The World of Underwater Micro-Grazers

Humans often forget that not so long ago, we lived in constant danger, threatened by predators. Today, our concerns have shifted to modern dangers like car accidents. However, in nature, all organisms feed and are consumed, rarely reaching old age. Thus, discussing predators is essential in this book.

Microscopic Predators of the Sea: Protozoans at Work

These single-celled powerhouses, with their complex nuclei, are the terrors of the oceanic microbial food web. They consume about 50–80% of the roughly 50 billion tons of carbon produced by microalgae each year, far more than their crustacean neighbors, the copepods.

Protozoans (Fig. 4.1) are not picky eaters. Their diet, depending on the species, includes algae, bacteria, fellow protozoans, and even creatures several times their size. But how do these minuscule organisms eat without mouths? Protozoans have evolved an array of feeding tactics, from filtration systems reminiscent of whales to engulfing large prey whole like

A. Calbet, *Plankton in a Changing World*,
https://doi.org/10.1007/978-3-031-76121-8_4

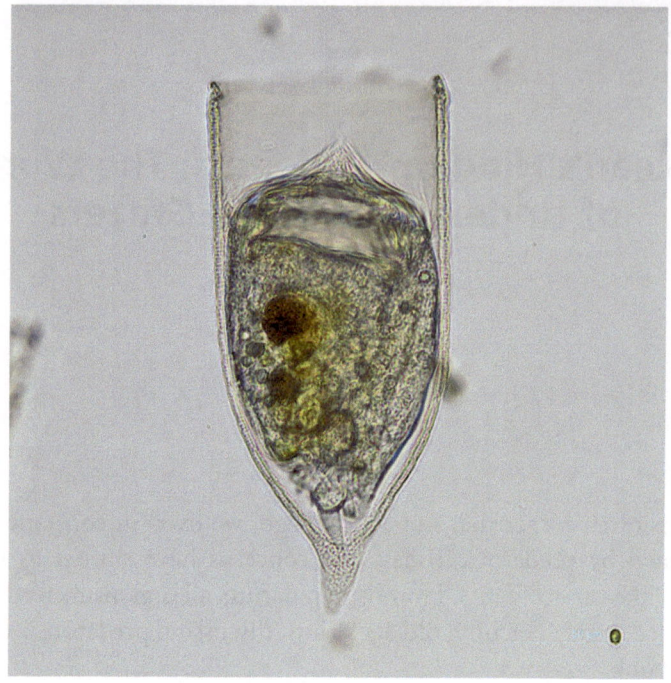

Fig. 4.1 *Favella* sp. Marine protozoan (tintinnid ciliate). © Albert Calbet

a snake. They deploy stinging toxins, vampiric tubes, and even external stomach-like structures (Fig. 4.2) to digest their catches. Thankfully, these microscopic marvels are not the size of sharks; otherwise, nobody would soak a foot in the water!

The Carnivorous Plants of the Ocean: The Story of Mixoplankton

Now, let's turn our attention to the ocean's carnivorous plants, the mixo-plankton (Fig. 4.3). These organisms may be tiny, but they defy the plant kingdom's usual rules by actively hunting prey. With a vast menu ranging

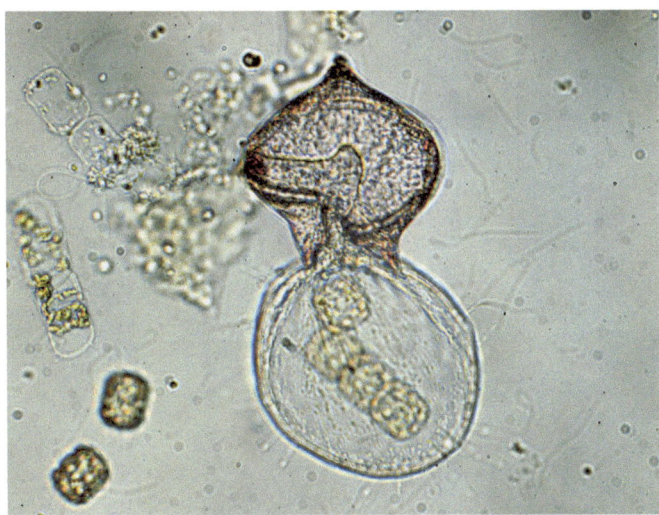

Fig. 4.2 A dinoflagellate (*Protoperidinium* sp.) pallium-feeding on a diatom chain. Notice the "stomach-like" structure, evaginated from the dinoflagellate cell, that trapped the diatoms. © Albert Calbet

from algae to multicellular animals, they trap and immobilize their targets using spears soaked in venom or by releasing potent toxins to the water. Among the carnivorous plants highlights a group that has no equivalent on land, the non-constitutive mixotrophs (Fig. 4.4): marine animals that have adapted to use the photosynthesis of enslave algae (or their chloroplasts); imagine a green rabbit photosynthesizing! It is an incredible world where the line between plant and animal blurs, and every species is a unique part of this complex puzzle. Investigating into the history of these organisms, we find that these marine protists likely played a pivotal role in the evolution of the very cells that make up all plant life. Mixotrophy in the ocean continues to amaze scientists, as each discovery leads to more questions than answers.

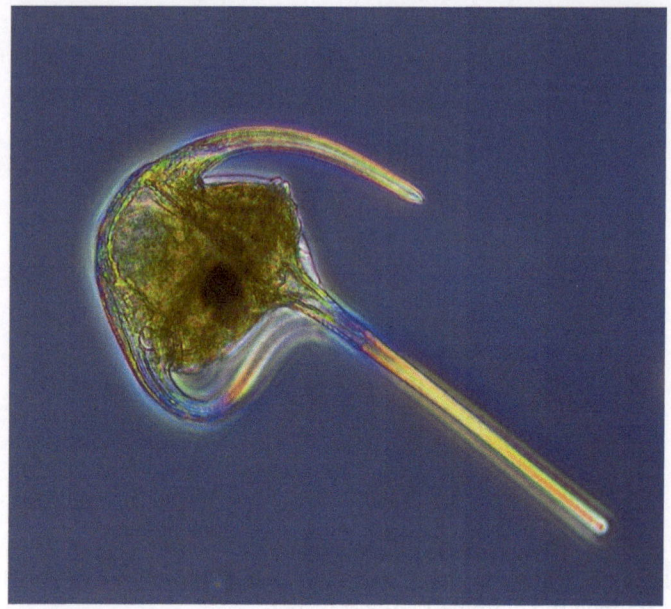

Fig. 4.3   Mixoplanktonic dinoflagellate. *Tripos gibberum.* © Albert Calbet

The Mighty Copepods: Tiny Titans Fueling Marine Life

Finally, let's present the copepods (Fig. 4.5), the miniature marvels of the sea. Despite their diminutive size, copepods are the ocean's (and the entire world) most populous metazoans, with an astounding display of 12,000 different species. These tiny crustaceans are everywhere, from free-swimming in the ocean to odd-shaped parasites living on diverse hosts.

Copepods are capable of swimming through the water at speeds that defy their size. Imagine being able to leap over skyscrapers in a single bound—that is the kind of speed we are talking about! Their daily vertical migrations are a testament to their endurance and importance in the ocean's food web. Equipped with specialized appendages, copepods are adept hunters and grazers, playing a crucial role in channeling nutrients from the micro to the macro realms. They are also the buffet of the sea, feeding fish larvae and sustaining commercial fisheries. But as climate

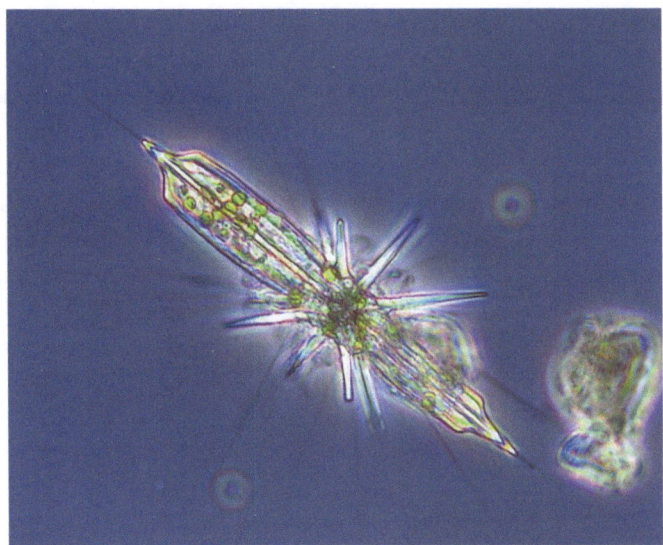

Fig. 4.4 A mixotrophic acantharean (*Amphibelone* sp.) hosting endosymbiotic algae (green dots). © Albert Calbet

Fig. 4.5 Marine copepod. *Centropages violaceus*. © Albert Calbet

change alters their distribution, the ripple effects could impact the entire marine ecosystem. In aquaculture, copepods are prized for their nutritional quality, including high levels of omega-3 and other essential nutrients, yet raising them in captivity poses a significant challenge.

In summary, the ocean's phytoplankton grazers, from the voracious protozoans to the predatory mixoplankton, and the indispensable copepods, are the hidden players of the aquatic world. Each group performs a critical role in maintaining the delicate balance of marine ecosystems.

5

Bacteria and Archaea: The Crucial Roles of Prokaryotes in Ocean Environments

In nature, the smallest and simplest an organism is, the most abundant and widespread becomes. Prokaryotes, which include bacteria and archaea (a primitive often very specialized group of prokaryotes), are ubiquitous and immensely diverse microorganisms playing fundamental roles in the seas and the ocean. These single-celled organisms lack a nucleus and other membrane-bound organelles, yet they thrive in a wide range of marine environments, from surface waters to the deepest ocean trenches. The diversity of marine prokaryotes is astounding, and their functions are crucial for maintaining the health and stability of marine ecosystems.

Bacteria: The Marine Workhorses

Marine prokaryotes are classified into two main groups: bacteria (Figs. 5.1 and 5.2) and archaea. Bacteria, the most studied of the two, are abundant in virtually all marine habitats. They are particularly prolific in the euphotic zone, where sunlight supports photosynthesis. Among these, cyanobacteria stand out. These photosynthetic bacteria are primary producers

A. Calbet, *Plankton in a Changing World*,
https://doi.org/10.1007/978-3-031-76121-8_5

Fig. 5.1 Epifluorescence microphotography of a natural microbial community stained with a DNA dye (DAPI), showing bacteria and some small eukaryotes. The bright yellowish tiny dots mostly correspond to heterotrophic bacteria and the red ones indicate the presence of photosynthetic organisms. © Albert Calbet

in the oligotrophic open ocean, converting carbon dioxide into organic matter using sunlight, like eukaryotic phytoplankton. This process not only forms the base of the marine food web but also significantly contributes to the global carbon cycle. Furthermore, cyanobacteria play a vital role in nitrogen fixation, converting atmospheric nitrogen into forms usable by other organisms, enriching the nutrient content of marine environments.

Archaea: The Hidden Specialists

Archaea, although less well-known, are equally important in marine ecosystems. Initially thought to be confined to extreme environments, such as hydrothermal vents and hypersaline lagoons, recent research has revealed that they are widespread and abundant even in more moderate marine habitats. Archaea are particularly notable for their roles in the

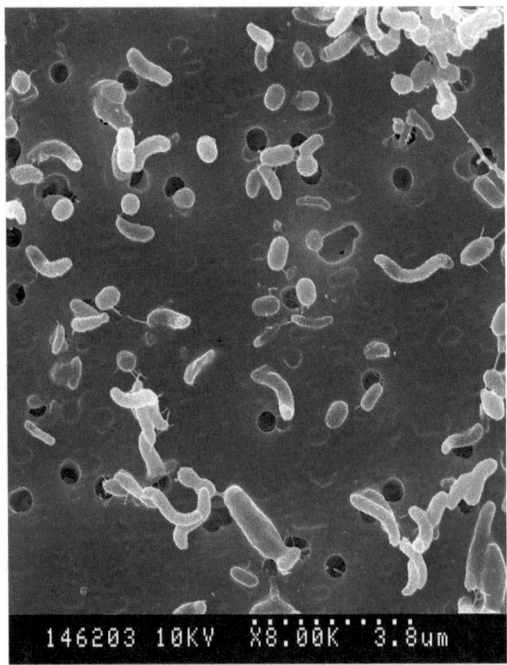

Fig. 5.2 Scanning electron microscopy of marine bacteria. © D. Vaqué and J.M. Fortuño

nitrogen and sulfur cycles. Ammonia-oxidizing archaea, for instance, are key players in nitrification, converting ammonia to nitrite and then to nitrate, essential nutrients for marine phytoplankton. Sulfur-oxidizing and sulfate-reducing archaea contribute to the sulfur cycle, influencing the chemistry of marine sediments and helping detoxify sulfur compounds.

Decomposers and Pollution Fighters

Prokaryotes play a critical role in the degradation of organic matter in the ocean. Heterotrophic bacteria, which obtain their energy from organic compounds, are responsible for the decomposition of dead organisms and organic waste. This decomposition process recycles nutrients back into the ecosystem, making them available for use by other organisms, such as phytoplankton. Additionally, some marine bacteria can break down complex pollutants, for example oil and plastics, playing a crucial role in mitigating pollution and maintaining marine health.

Symbiotic Relationships

Symbiotic relationships between prokaryotes and other marine organisms are another fascinating aspect of their ecological roles. Many marine animals, including corals, sponges, and deep-sea vent animals, host symbiotic bacteria that provide them with essential nutrients or aid in metabolic processes. For example, certain bacteria living within the tissues of deep-sea vent organisms can oxidize hydrogen sulfide, providing their hosts with energy in an otherwise nutrient-poor environment. Similarly, nitrogen-fixing bacteria in the tissues of some marine plants and animals supply them with bioavailable nitrogen, a critical nutrient in many marine habitats.

Moreover, bacteria in the digestive tracks of marine organisms, as it occurs in the terrestrial ones, play crucial roles in digestion, nutrient synthesis, pathogen protection, and detoxification. They help break down complex organic materials, enabling efficient nutrient absorption, especially in nutrient-poor environments. These bacteria synthesize essential nutrients like vitamins and amino acids, supporting their host's metabolic functions. Additionally, they act as a biological barrier against pathogens by outcompeting harmful microorganisms and producing antimicrobial compounds. Gut microbiota also metabolize environmental toxins, reducing their toxicity and aiding in excretion. Their presence ensures gut health by maintaining the integrity of the gut lining, preventing harmful substances from entering the bloodstream. This symbiotic relationship highlights the mutual benefits and interdependence between marine organisms and their gut bacteria, underscoring the bacteria's indispensable role in the health and survival of marine life.

Climate Regulators

Marine prokaryotes also influence the Earth's climate. Certain microbial activities produce methane, a potent greenhouse gas, and its production and release from marine sediments and the water column can significantly impact global warming. Conversely, the DMS produced by the breakdown of dimethylsulfoniopropionate (DMSP) by marine bacteria can influence cloud formation and potentially cool the Earth's surface by reflecting sunlight back into space. Thus, the metabolic activities of marine prokaryotes have far-reaching implications for global climate regulation.

Unveiling the Hidden World

The study of marine prokaryotes is still evolving, and advanced techniques such as metagenomics and single-cell sequencing are uncovering new species and functions at an unprecedented rate. These technologies allow scientists to explore the genetic diversity of prokaryotic communities and understand their roles in marine ecosystems more comprehensively. As research progresses, it becomes increasingly clear that prokaryotes are indispensable to the health and functioning of the ocean, influencing everything from nutrient cycling and pollution degradation to symbiotic relationships and climate regulation.

6

Tiny Terrors of the Ocean: Planktonic Viruses and Parasites

Being eaten by a predator is not the only way plankton may perish. Viral and parasitic attacks are also common causes of death for many marine creatures. Viruses and parasites are just as diverse as their free-living counterparts and play crucial roles in shaping ocean health. Their impact extends to the very foundation of our seafood industry: fisheries and aquaculture. This chapter explores their fascinating and often lethal interactions with other organisms.

Viruses: Minuscule but Efficient Killers

For decades, the underwater world of plankton was primarily understood through the lens of bacteria and larger groups. However, a revolutionary discovery in the latter half of the twentieth century revealed the existence of virioplankton (planktonic viruses; Figs. 6.1 and 6.2), a previously hidden world of abundant viruses silently shaping the dynamics and structure of aquatic ecosystems.

Imagine a single drop of ocean water. While it might appear crystal clear to the naked eye, it is teeming with life, including a shocking ten

© The Author(s), under exclusive license to Springer Nature Switzerland AG 2024
A. Calbet, *Plankton in a Changing World*,
https://doi.org/10.1007/978-3-031-76121-8_6

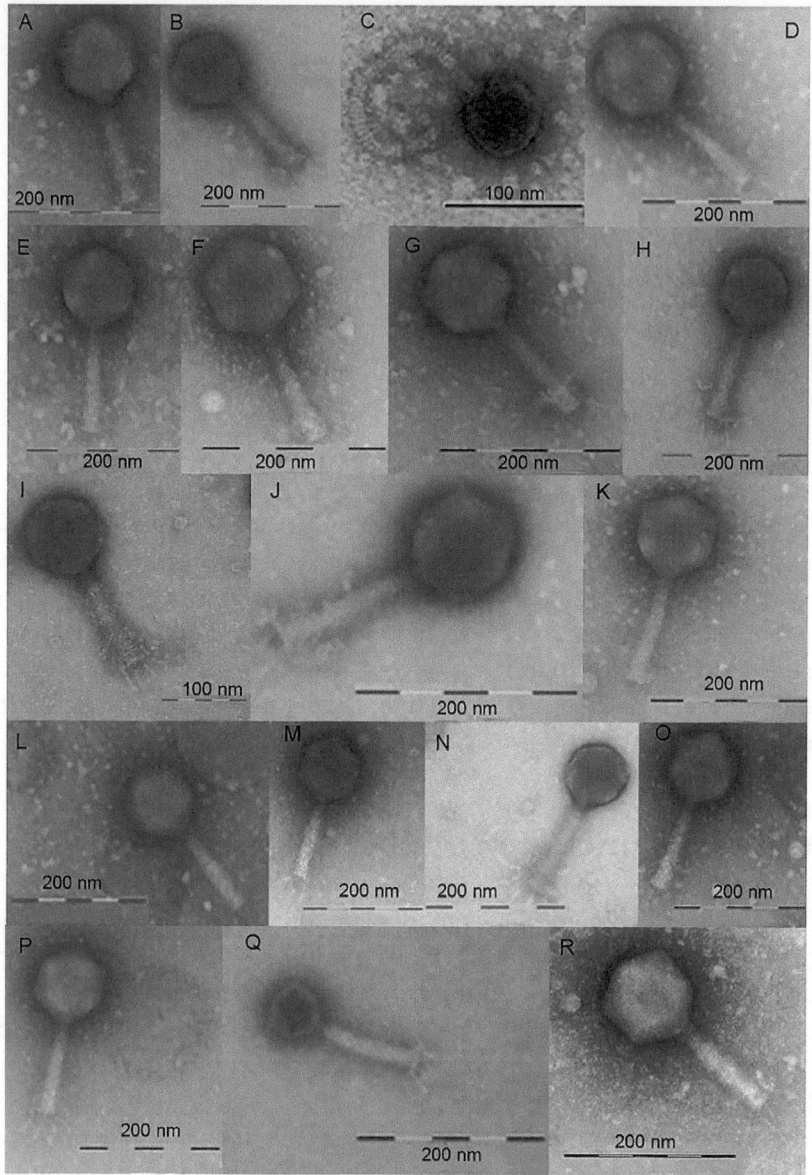

Fig. 6.1 Electron microscopy images of marine viruses. nm = nanometer (0.000001 mm). © D. Vaqué and E. Lara

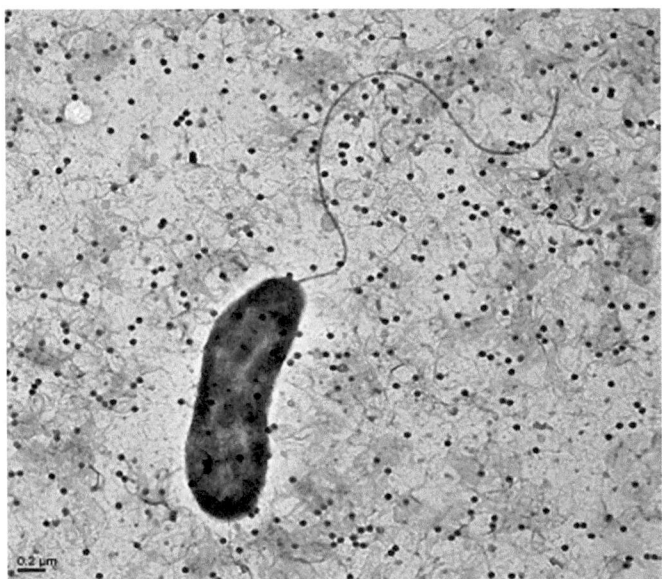

Fig. 6.2   Electron microscopy images of a marine bacteria (*Pseudoalteromonas* sp.) and marine viruses in the background (dark dots). Note the size differences between bacteria and viruses. © D. Vaqué and E. Lara

times more viruses than bacteria!—just for your info, bacteria may reach common abundances of 1,000,000 individuals per ml in coastal waters. This abundance hints at the significant, yet often unnoticed, role viruses play in aquatic environments.

Unlike their more static counterparts, virioplankton populations are dynamically fluctuating, exhibiting significant variations in abundance within short timeframes (e.g., minutes or hours). Typically, their numbers surge during periods of high biological activity, like algal blooms. This dynamic nature highlights the complex interplay between environmental factors, the availability of host organisms for viruses to infect, and the rapid replication cycles of these microscopic entities.

Marine viruses exhibit several lifecycles, each influencing how they interact with their hosts and impact surrounding communities. Here are some examples:

Lytic Cycle: The virus infects a host cell, replicates its genetic material within the host, and then bursts the cell (lysis) to release new viral particles, ultimately killing the host. This rapid replication and lysis can lead to significant and fast mortality rates in host populations.

Lysogenic Cycle: The viral genetic material integrates into the host's genome, becoming a prophage. The prophage can lie dormant for generations, replicating along with the host cell. Under certain environmental conditions, the prophage can be activated, switching to the lytic cycle and lysing the host cell. This allows the virus to persist within a host population even during periods of low abundance.

Pseudolysogenic Cycle: Similar to the lysogenic cycle, but instead of integrating its entire genome, the virus integrates smaller portions, allowing them to enter a standby mode when conditions are unfavorable. When conditions improve, they can switch back to more active replication cycles.

Chronic Cycle: Chronic phages establish a long-term relationship with the host bacteria. These phages bud out of the bacterial membrane in a controlled manner, without killing the host cell. This relationship is like a partnership, where phages help bacteria form biofilms and transfer genes that might make the bacteria more harmful, helping them in toxin production.

Beyond their sheer numbers and dynamic presence, virioplankton play a crucial role in the flow of organic matter within aquatic ecosystems. Through a process called the "viral shunt," viruses infect and lyse (destroy) their host cells, releasing trapped nutrients and carbon as dissolved organic matter (DOM). This process diverts a significant portion of organic carbon, which would otherwise be directly consumed by higher trophic levels like zooplankton, toward the DOM pool.

Quantifying the exact contribution of viruses to this process is challenging, but estimates suggest they contribute to the daily loss of around 20% of heterotrophic bacteria and 3–5% of phytoplankton cells. This translates to a substantial impact on energy flow and nutrient cycling within aquatic ecosystems. Moreover, some researchers believe viruses might act as selective agents, preferentially targeting dominant or fast-growing strains of bacteria or phytoplankton. By doing so, they could indirectly influence the composition and diversity of these communities, maintaining a more balanced ecosystem (Kill the Winner hypothesis, Chap. 14).

Parasites: The Merciless Invaders

Parasitic plankton come in various forms, often adopting bizarre shapes very different from their free-living relatives. They range from unicellular to multicellular organisms, showcasing remarkable predatory strategies. For instance, the infamous sea lice, which belong to the copepod group, attach to salmon and other fish species, feeding on their skin and blood (Fig. 6.3). These parasitic copepods harm fish and cause significant economic losses in the aquaculture industry. In Norway alone, sea lice infestations are estimated to cost the salmon farming industry hundreds of millions of dollars annually. Interestingly, copepods themselves are often invaded by parasitic dinoflagellates, such as *Blastodinium*, or even by other crustaceans (Fig. 6.4).

Another intriguing example is the parasitic chytrid fungus. This microscopic predator burrows into the cell walls of some microalgae, a vital

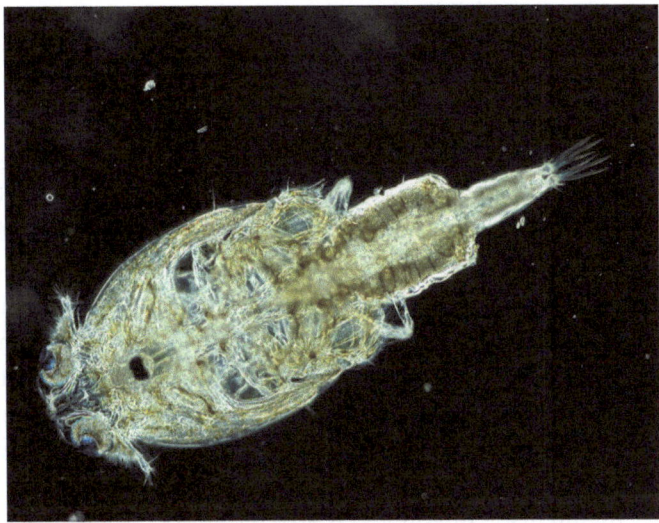

Fig. 6.3 Parasitic copepod of the sea lice family. © Albert Calbet

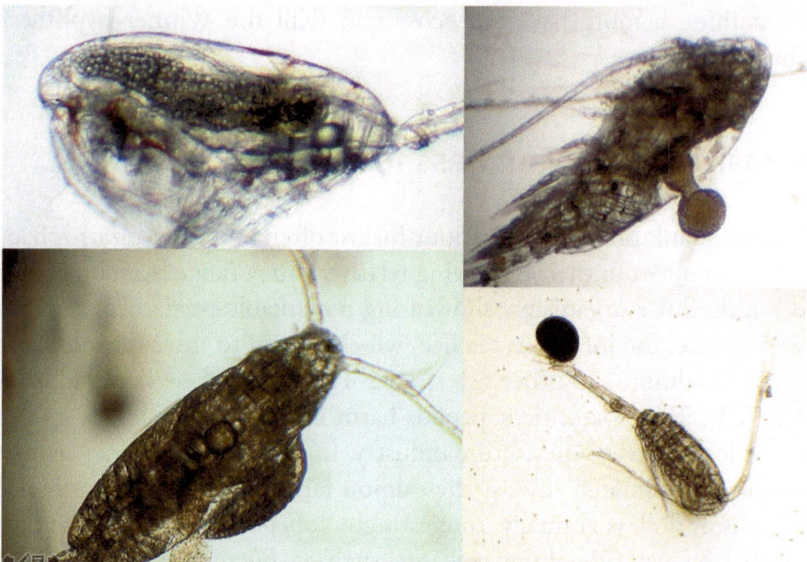

Fig. 6.4 Examples of copepods infected by parasites. © Albert Calbet. Top left: *Blastodinium* sp.; top right: *Ellobiopsis* sp.; bottom left: parasitic isopod; bottom right: *Ellobiopsis* sp.

food source for many fish larvae and zooplankton. By infecting and ultimately killing these microalgae, parasitic chytrids can indirectly impact the survival and growth of commercially important fish populations in fisheries.

Parvilucifera is a genus of microscopic parasitic protists, specifically targeting dinoflagellates, some of which can form harmful algal blooms. These diverse predators, found worldwide, have a complex lifecycle involving free-swimming infective stages, intracellular development within the host, and release of new infective spores. Their sophisticated strategies include amoeboid movement for host searching, enzyme secretion for cell wall penetration, and even toxin production to manipulate host behavior. By controlling the populations of their hosts, including harmful bloom species, *Parvilucifera* plays a crucial role in maintaining a balanced marine ecosystem. Their potential to control harmful algal blooms highlights ongoing research into these fascinating micro-predators.

Implications for a Sustainable Future

The knowledge gained from studying viruses and parasitic plankton is crucial for informing sustainable practices in fisheries and aquaculture. By understanding how these tiny predators influence fish populations and the broader ecosystem, we can develop strategies to minimize their negative impacts while fostering their potential benefits. This could involve implementing stricter regulations on aquaculture practices to reduce the spread of sea lice or exploring the potential of using certain parasitic plankton as a natural biocontrol method against harmful algal blooms.

7

The Hidden World of Marine Fungi

Have you ever gone mushroom hunting in the forest? It is quite fun, and if you can differentiate edible species from toxic ones, you might end up with a delicious dinner!—otherwise, you may end up at the hospital.

Fungi have numerous applications, including food, medicine, agriculture, and biotechnology. What most people do not realize is that fungi also thrive in the ocean, playing a vital role in its complex ecosystems. Found in diverse habitats, ranging from coastal mangroves to the deep sea, marine fungi are integral to nutrient cycling, organic matter decomposition, and symbiotic relationships with marine plants and animals. Despite their ecological significance, marine fungi have remained relatively understudied compared to their terrestrial counterparts. This lack of attention is gradually shifting as researchers uncover the many ways these organisms contribute to ocean health and their potential applications in biotechnology and environmental management.

A. Calbet, *Plankton in a Changing World*,
https://doi.org/10.1007/978-3-031-76121-8_7

The Astonishing Diversity of Marine Fungi

The diversity of marine fungi is astounding, with thousands of species identified, each adapted to the unique conditions of their habitats. These fungi can be found free-living in the plankton, on submerged wood, in sediments, and within marine animals and plants, displaying a range of physiological adaptations that allow them to thrive in saline, often extreme environments. For instance, marine fungi have developed specialized enzymes to break down complex organic materials, facilitating the recycling of nutrients in marine ecosystems. This capability is crucial for maintaining healthy oceanic environments, ensuring the continuous turnover of organic matter and supporting the food web from the smallest microorganisms to the largest marine mammals.

Fungal Parasites

Marine fungi also act as parasites, infiltrating and breaking down phytoplankton cells (Fig. 7.1). This fungal attack not only regulates phytoplankton populations but also makes the complex sugars within the algae accessible to other decomposers, particularly bacteria. This intricate dance of consumption and breakdown is fundamental to the ocean's carbon cycle, ensuring a continuous flow of carbon from one organism to the next, ultimately returning to the atmosphere as carbon dioxide.

Decomposition and Pollution Cleanup

The ocean floor is another stage where marine fungi play a starring role. Here, they act as saprotrophs, consuming dead and decaying organic matter that settles on the seabed. This tireless process of decomposition is essential for releasing trapped nutrients back into the water column, making them available for phytoplankton and other marine life. Marine fungi even play a role in cleaning up oil spills. Their unique enzymatic

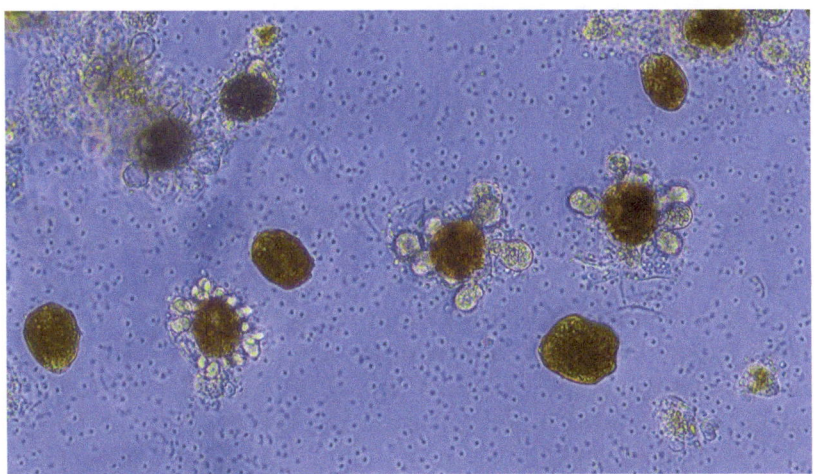

Fig. 7.1 Marine fungi (*Dinomices arenysensis*) infecting dinoflagellates (*Ostreopsis* sp.) © Albert Calbet, from a culture provided by E. Garcés

machinery allows them to break down complex hydrocarbons, potentially aiding in the bioremediation of polluted marine environments.

Biodegrading Plastics: A New Hope

One of the most exciting recent discoveries in marine mycology is the ability of certain marine fungi, such as *Parengyodontium album*, to degrade plastics. This breakthrough has significant implications for addressing marine pollution. This particular species of marine fungus can decompose polyethylene, one of the most common and persistent types of plastic found in the ocean. The process is facilitated when the plastic has been exposed to ultraviolet (UV) radiation from sunlight, which alters the plastic's chemical structure and makes it more susceptible to fungal degradation. This discovery highlights the potential for biotechnological solutions to mitigate the pervasive problem of plastic pollution, opening new avenues for waste management and underscoring the importance of protecting and studying these organisms within their natural habitats.

Pharmaceuticals from the Ocean

In addition to their ecological roles, marine fungi are a treasure trove of bioactive compounds with promising pharmaceutical applications. Marine fungi produce a wide array of secondary metabolites, including antibiotics, antifungals, and anticancer agents, many of which are unique to marine environments. The extreme conditions under which these fungi live often lead to the production of novel chemical structures not found in terrestrial organisms. Researchers are increasingly turning to marine fungi in the quest for new drugs, driven by the urgent need for novel treatments against resistant pathogens and other diseases. The exploration of marine fungal metabolites is still in its early stages but holds immense potential for medical science and biotechnology.

Symbiotic Relationships: Enhancing Marine Plant Health

Marine fungi are integral to the health of marine plants, forming symbiotic relationships that enhance plant growth and resilience. For example, many marine fungi establish mutualistic associations with seagrasses and algae, aiding in nutrient absorption and providing protection against pathogens. These interactions are critical for the stability of coastal ecosystems, which are among the most productive and biodiverse regions on the planet. Understanding the dynamics of these relationships is essential for the conservation and restoration of vital marine habitats, especially in the face of climate change and human activities that threaten these ecosystems.

Part II

Plankton Distribution and Habitats

8

The Biomass Distribution in Earth's Ecosystems

Humans often compare things: Is this better than that? Is that one taller than me? In this chapter, we will extend this habit to comparing ecosystems, particularly their biomass distribution. First, we need to define the rules. The distribution of living organisms on Earth can be quantified using various units, such as the number of individuals or species. However, biologists often use a different unit: biomass. Biomass refers to the total mass of living organisms in a given area or ecosystem, encompassing everything from microscopic viruses to gigantic trees, as well as the organic matter they produce or consume. This unit is particularly useful for understanding the contribution of different groups of organisms to the fluxes of elements, often described in carbon units such as Gt C (Gigatons of carbon = 1,000,000,000,000 kg) for global estimates.

Within the different ecosystems on Earth, the biomass pyramid serves as a fundamental representation of energy flow and trophic structure. Comparing the biomass pyramid between marine and terrestrial ecosystems reveals fascinating differences shaped by the unique challenges and opportunities each environment presents.

© The Author(s), under exclusive license to Springer Nature Switzerland AG 2024
A. Calbet, *Plankton in a Changing World*,
https://doi.org/10.1007/978-3-031-76121-8_8

The Classic Pyramid: Terrestrial Biomass Distribution

The terrestrial biomass pyramid (Fig. 8.1), often depicted in textbooks, resembles an ancient Egyptian or American pyramid with a broad base and a narrowing apex. This shape reflects the basic principle of energy transfer in ecosystems, known as the 10% rule of thumb. As energy flows through the food web, approximately 90% is lost as heat at each trophic level due to cellular respiration and other processes. Consequently, the producers, primarily plants, form the vast base of the pyramid, supporting a smaller population of herbivores (consumers) at the next level. These herbivores, in turn, support an even smaller population of carnivores (higher-order consumers) at the top. This progressive decrease in biomass as we move up the food web results in the characteristic pyramid shape.

However, it is important to note that this is a simplified model, and real-world ecosystems exhibit more complexity. For example, decomposers play a vital role in breaking down dead organic matter and returning nutrients to the system, forming a detritus food web that runs parallel to the grazing food web. Additionally, some ecosystems may have multiple trophic levels within each consumer category, further impacting the pyramid shape.

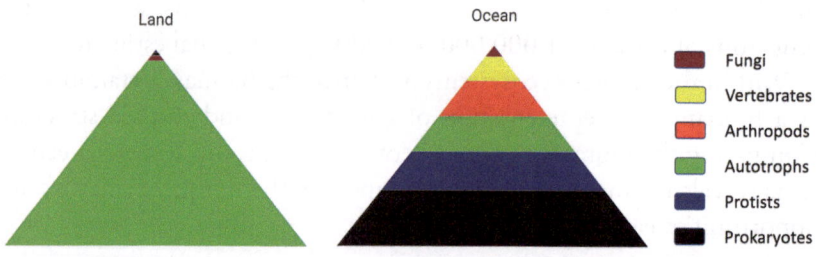

Fig. 8.1 Terrestrial (left) and marine (right) major-groups biomass pyramids. Note that autotrophs do not conform the base of the pyramid in marine systems. If a biomass with autotroph at the base would have been generated, this would have a diamond-like shape. Data for the graph obtained from https://www. encyclopedie-environnement.org/en/life/distribution-biomass-planet/

The Marine Biomass Rhomboidal or Inverted Pyramid

Marine ecosystems present a different picture. Unlike the terrestrial pyramid, the marine biomass pyramid often resembles an inverted pyramid (Fig. 8.1) or a diamond shape. This unique structure reflects the particular contribution of microscopic organisms like phytoplankton and bacteria to the total biomass. Despite their small size and little biomass, primary producers (phytoplankton, Fig. 8.2) form the base of the food web, supporting a diverse range of zooplankton and other consumers. Indeed, phytoplankton present a very low biomass compared to terrestrial plants (1 vs 450–470 Gt C, respectively), but they boast similar production rates (estimated at around 50 Gt C per year for both), owing to the rapid reproduction and turnover rate of phytoplankton. Thus, while terrestrial plants (Fig. 8.3) hold the crown in terms of total mass, marine phytoplankton wear the production power belt, highlighting the diverse and complex nature of life on Earth, where size does not always dictate efficiency.

As we move up the food web, the biomass of consumers increases. Zooplankton feed on phytoplankton and have an estimated biomass similar to that of phytoplankton. Further up the food web, fish (including

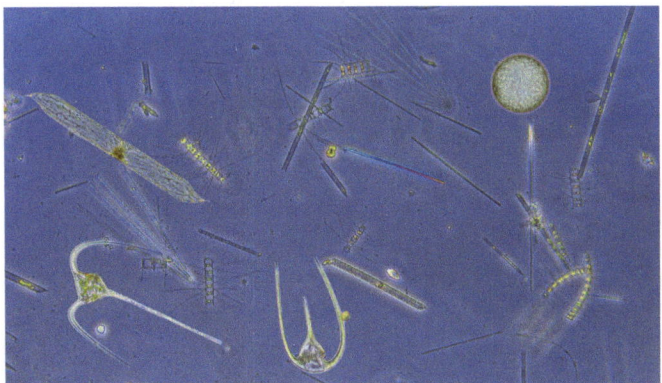

Fig. 8.2 Marine phytoplankton. NW Mediterranean coastal waters. © Albert Calbet

Fig. 8.3 Terrestrial plants have a much larger biomass than their marine counterparts. However, both support similar global photosynthesis. Kawaii Island (Hawaii, USA) © Albert Calbet

small fish, forage fish, and larger predatory fish) have an estimated biomass of approximately 0.7 Gt C. This cumulative biomass of predators reflects the accumulation of energy as it moves up the food web. Large marine mammals, like whales, and top predators, like sharks, sit at the top of the food web and, despite their impressive size, have a relatively low estimated biomass, approximately 0.004 Gt C. This is because they are apex predators with smaller populations and longer lifespans, requiring less food in total compared to the lower trophic levels. Just for curiosity, humans represent 0.06 Gt C, although the total biomass of crops cultivated by humans is estimated at approximately 10 Gt C.

Biomass Distribution Beneath Our Feet

Deep beneath the surface of our planet lies a hidden world teeming with life—a vast biosphere dominated by bacteria and archaea. These microscopic prokaryotes, often overlooked and underestimated, hold the title of one of the most abundant life forms on Earth, residing in staggering numbers not only in our familiar environments but also in the depths of the terrestrial and marine subsurface. Estimates suggest that bacteria and archaea collectively contribute a staggering 30–60 Gt of carbon to Earth's total biomass, accounting for roughly 70% of the biomass in the deep terrestrial subsurface. This overshadows the biomass of all animals on our planet, highlighting the sheer scale of this microbial domain. For comparison, the combined biomass of bacteria and archaea in surface soil and marine ecosystems accounts for 7.5 and 1.6 Gt C, respectively. The vast biomass of bacteria and archaea in the subsurface plays a crucial role in various biogeochemical cycles, influencing processes like carbon sequestration, nutrient cycling, and methane production.

Moreover, the subsurface environment presents a unique set of challenges for life. Sunlight, the primary energy source for most organisms, is absent, forcing bacteria and archaea to rely on alternative energy sources. They can utilize chemical energy from various processes, such as the breakdown of minerals or the oxidation of methane, to fuel their growth and survival. Additionally, the subsurface environment is often devoid of oxygen, necessitating the dominance of anaerobic organisms—those that can thrive in the absence of oxygen. These adaptations allow bacteria and archaea to flourish in this challenging realm, forming the cornerstone of the hidden biosphere.

Comparing the shapes of terrestrial and marine biomass pyramids reveals the fascinating diversity of how life organizes itself on Earth. The terrestrial pyramid reflects the principle of diminishing energy as it progresses through the food web, while the marine pyramid showcases the significant role of microscopic organisms and their fast cycling of matter and energy. Understanding these differences is crucial for appreciating the unique characteristics of each ecosystem and for developing effective conservation strategies to protect the delicate balance of life on our planet.

9

A Patchy Ocean: From Microscopic Meals to Macroscale Aggregations

The subject of patchiness in the ocean evokes memories of my youth when I was conducting my Ph.D. on the effects of patchiness and other sources of variability on zooplankton production. The truth is, the ocean may appear uniform at first glance, but beneath the surface lies a world teeming with diversity and patchy distribution. This patchiness extends not only to the countless species of marine life but also to the very foundation of the marine food web: organic matter. Far from being a homogenous soup, marine organic matter exhibits remarkable heterogeneity at all scales, with significant variations in composition and availability occurring over distances ranging from mere micrometers to kilometers. This patchiness, from the microscopic to the macroscopic, plays a critical role in shaping the distribution and activity of marine life, from the invisible engines of the ocean, microbes, to the majestic creatures of the plankton world.

The Microscopic Buffet

At the microscale, there is a huge variety of tiny patches and gradients orchestrated by the laws of physics. Dissolved organic matter (DOM), released by phytoplankton and zooplankton excretion, sloppy feeding, or decaying organisms, constantly undergoes diffusion, which allows its molecules to spread out and encounter microorganisms throughout the water column. The rate of diffusion depends on the size and shape of the molecule, with smaller sugars and amino acids diffusing more readily than complex carbohydrates or lipids. This creates a dynamic patchwork of DOM concentration, with localized gradients guiding microbes toward areas rich in potential food sources.

A simple sinking phytoplankton cell releases a burst of organic carbon, creating a localized patch rich in readily available sugars and amino acids. Nearby bacteria, sensing the chemical gradient through specialized receptors, swarm toward this feast. As they consume the readily available nutrients, the composition of the organic matter patch changes. Microbes with specific enzymatic capabilities thrive in these patches, while others may struggle to find suitable food sources. This selective feeding behavior further contributes to the heterogeneity of organic matter at the microscale.

Fluid dynamics also play a crucial role in these microscopic interactions. Tiny eddies and turbulence arise due to water movements. These micro-eddies act like invisible mixers, stirring and mixing DOM and microbial communities, promoting encounters between microbes and their preferred food sources. Conversely, in calm environments, diffusion becomes the dominant force. Here, localized depletion zones can form around actively feeding microbes, as they rapidly consume the surrounding DOM, creating a temporary "food desert" until fresh DOM diffuses in.

Adhesion's Sticky Grip

Organic matter often adheres to surfaces, including sinking particles, marine snow aggregates (clumps of organic matter formed by colliding particles, Fig. 9.1), and even plankton. These interactions can

Fig. 9.1 Underwater scene with multiple aggregated particles forming marine snow, and a silhouette of a large manta ray at the bottom. Open water off Komodo Island. © Albert Calbet

significantly influence the fate of organic matter by slowing/accelerating its sinking rate or bringing it into close proximity with specific microbial communities. Certain bacteria possess adhesive surface features that allow them to attach to sinking particles or marine snow. This attachment provides them with a steady supply of organic matter as the particle descends, while also potentially slowing the sinking rate and allowing for further microbial degradation in the water column. Additionally, some microbes specialize in degrading specific types of organic material. For example, some bacteria excel at breaking down complex carbohydrates found on the chitinous exoskeletons of zooplankton, while others may be better equipped to handle the lipids found in fecal pellets.

The Influence of Light

Light penetration plays a critical role in shaping the distribution of organic matter at both the micro and macroscale. The depth to which light penetrates varies depending on the wavelength and water clarity.

This creates a vertical gradient of light availability, influencing the distribution of phytoplankton and the organic matter they produce. Phytoplankton communities are most abundant near the surface, where light is plentiful, and their contribution to organic matter production diminishes rapidly with depth. However, in the oligotrophic open ocean, where waters are clear and light penetrates very deep in the water column, we usually find the maximum abundance of phytoplankton at the base of the mixed layer, at 50-100 m depth. At these depths, there is still some light and nutrients can be replenished from below the thermocline. This patchy distribution of phytoplankton biomass, driven by the interplay of light and nutrients, translates into a corresponding patchiness of freshly produced organic matter, further influencing the distribution and activity of microbial communities throughout the water column.

Examples of Patchiness in the Ocean

The intricate dance between physical forces, organic matter distribution, and microbial communities at the microscale does not occur in isolation. These microscale processes play a fundamental role in shaping the larger-scale patterns of organic matter distribution in the ocean. Some examples of patches are:

Microbial Gliders: Certain bacterial species exhibit chemotaxis, allowing them to move toward areas with higher concentrations of preferred nutrients. This targeted movement can further enhance the patchiness of bacterial communities at the microscale.

Sinking Particles: Dead organisms, fecal pellets, and other particulate organic matter (POM) continuously rain down through the water column. The sinking rate of these particles depends on their size, density, and the viscosity of the water. Physical factors like turbulence can also influence this sinking process, creating patchy distributions of POM at different depths.

Marine Snow: Aggregates of marine snow, consisting of a mix of organic particles, can form as sinking POM collides in the water column. These marine snow "storms" can rapidly deliver organic matter to the deep

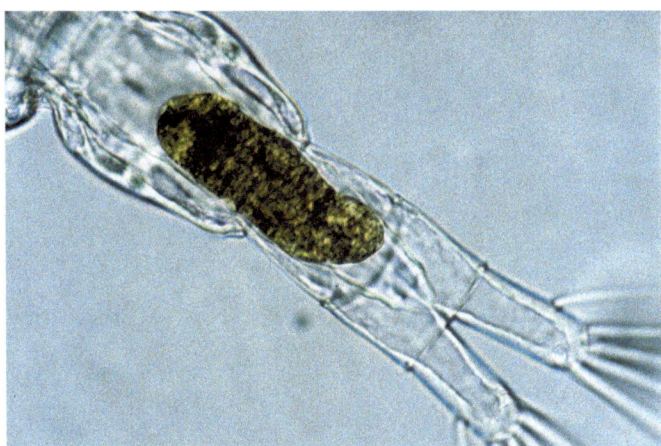

Fig. 9.2 Fecal pellet inside a copepod abdomen. © Albert Calbet

ocean, influencing the distribution of nutrients and fueling microbial communities in these dark depths.

Fecal Pellets: As zooplankton graze on phytoplankton, they sink, leaving behind fecal pellets (Fig. 9.2) rich in lipids and other organic material. These pellets create localized hotspots for bacteria with specialized enzymes for degrading these more complex molecules.

Harmful Algal Blooms: Certain phytoplankton species can produce toxins harmful to other marine life. When environmental conditions favor these toxin-producing species, they can form localized blooms, popularly known as red tides. These blooms create patchy areas of high toxin concentrations, posing a threat to other marine organisms, including zooplankton grazers.

Upwelling Zones: When deep-sea currents rise to the surface, they bring with them nutrient-rich water. This mechanism triggers localized blooms of phytoplankton, creating vast green patches visible from space. These blooms attract zooplankton, leading to further enrichment of the area.

Zooplankton Patchiness: Zooplankton often exhibit patchy distributions. Some species may aggregate in response to chemical cues released by phytoplankton or other prey, creating localized feeding hotspots. Others may exhibit diel vertical migration, moving up toward the surface at night to feed and returning to deeper waters during the day to avoid

Fig. 9.3 *Rhopilema esculenta*—Flame jellyfish. © Albert Calbet

predators. This vertical migration can create distinct layers of zooplankton abundance at different depths.

Oceanic Fronts: Fronts form where currents of different temperatures or salinities meet. These fronts can act as physical barriers, trapping plankton within a concentrated zone. This creates a rich feeding ground for higher trophic levels like fish. Zooplankton may also be selectively trapped at fronts based on their swimming abilities or tolerance for different water conditions.

Jellyfish and Other Zooplankton Blooms: Certain jellyfish (Fig. 9.3) and other zooplankton species can form massive aggregations at the surface, creating a dense patch of biomass. These aggregations can be driven by favorable environmental conditions or by the presence of abundant prey.

These are just a few examples, and the specific characteristics of these patches will vary depending on factors like the type of plankton, nutrient availability, and physical oceanographic processes.

10

Plankton Across Ecosystems

If you have traveled around the world, you may have noticed that the waters of different coastal areas have distinct colors. This chromatic palette (usually from deep blue to green and brown) may be due to solids in suspension, but it is quite likely influenced by the plankton inhabiting those waters. Plankton vary greatly depending on their ecosystem, adapting to the unique conditions of polar, tropical, coastal, open ocean environments, and different ocean depths. Understanding these differences not only sheds light on the diversity of plankton but also highlights their crucial roles in maintaining the balance of marine ecosystems globally. This chapter explores the fascinating variations of plankton across these diverse environments, providing rich examples and insights along the way.

From Polar Regions to Tropical Waters

Polar regions, both Arctic and Antarctic, present some of the most extreme and challenging environments on Earth. The waters here are characterized by low temperatures, seasonal ice cover, and unique light conditions, which influence the types of plankton that thrive.

A. Calbet, *Plankton in a Changing World*,
https://doi.org/10.1007/978-3-031-76121-8_10

In the Arctic, phytoplankton such as diatoms dominate during the short summer months when the ice melts and sunlight penetrates the water. Diatoms can rapidly bloom, creating large expanses of greenish-colored water visible even from space. These blooms are crucial for supporting the entire Arctic food web, from tiny zooplankton like copepods to large marine mammals such as whales. The Antarctic region (Fig. 10.1) hosts a similar abundance of diatoms, but the waters also harbor unique species that can survive within the ice. These ice-associated microorganisms have adapted to extreme salinity and cold, capable of photosynthesis at very low light levels. Antarctic krill, a critical component of the Southern Ocean ecosystem, rely heavily on these plankton. During winter, krill graze on the algae growing on the underside of sea ice. Other zooplankton groups in polar regions also exhibit remarkable adaptations. For instance, many large Arctic copepods of the genus *Calanus* have life-cycles synchronized with the seasonal availability of phytoplankton. They accumulate lipid reserves during summer blooms, which help them survive the long, dark winters. These adaptations underscore the resilience and specialization of polar plankton, which have evolved to exploit the fleeting productivity of their icy habitats.

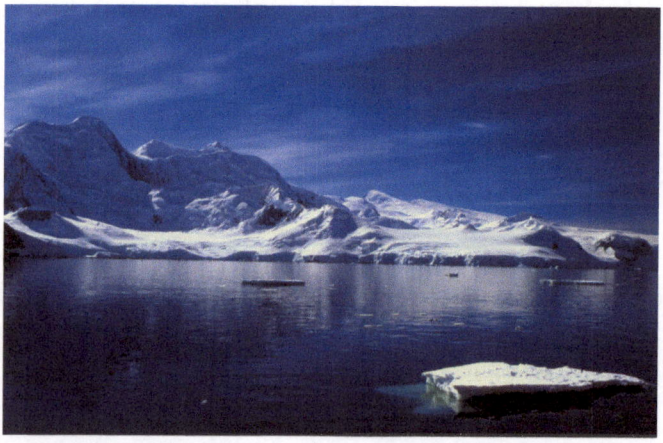

Fig. 10.1 The Gerlache Strait, Antarctica. © Albert Calbet

Shifting to tropical waters, the scenario changes drastically. These warm, sunny regions are teeming with an astonishing variety of tiny plankton, driven by stable temperature conditions and abundant solar energy year-round. However, nutrient availability in tropical waters can be highly variable, often leading to nutrient limitation. In tropical regions, cyanobacteria, specifically *Prochlorococcus* and *Synechococcus*, play a pivotal role. *Prochlorococcus* is perhaps the smallest known photosynthetic organism (less than 1 μm in diameter) and dominates the open tropical ocean, driving much of the primary production. Its ability to thrive in low-nutrient environments makes it crucial for carbon cycling and contributes significantly to global oxygen production. Zooplankton in tropical ecosystems showcase a wide array of forms and behaviors. The wide diversity of zooplankton supports the rich fish populations in coral reefs, highlighting the interconnectedness of tropical marine ecosystems.

From Coastal to Open Sea

Coastal ecosystems, where land and sea interact, offer unique environmental conditions that support diverse and productive plankton communities. Nutrient inputs from rivers, estuaries, and upwelling zones create fertile grounds for phytoplankton growth, leading to higher biomass compared to the open ocean. Diatoms are particularly abundant in coastal waters, benefiting from nutrient-rich conditions. Genus like *Skeletonema* and *Chaetoceros* thrive in these environments, often forming large blooms. These blooms serve as a vital food source for many marine organisms, fueling coastal food webs. Additionally, coastal areas are hotspots for other groups like coccolithophores—tiny phytoplankton that produce calcium carbonate plates—contributing to the marine carbon cycle and sediment formation. Coastal waters are also home to unique zooplankton species. Various copepods, such as *Centropages typicus* (Fig. 10.2), *Eurytemora affinis*, and different species of the *Acartia* genus are abundant and form key links between phytoplankton and higher trophic levels. These transitional regions act as nurseries for many fish and invertebrate species, highlighting the importance of plankton in

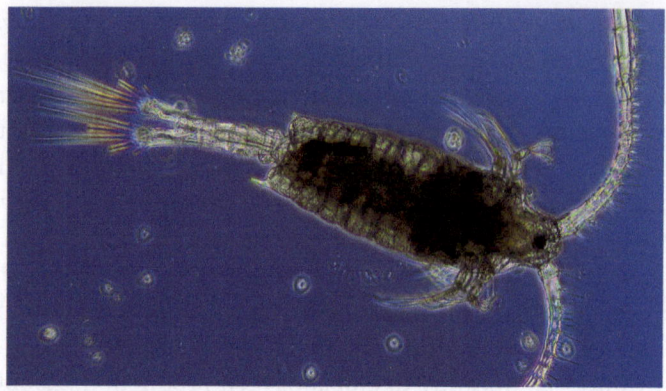

Fig. 10.2 *Centropages typicus.* © Albert Calbet

supporting larval development and growth. Moreover, coastal ecosystems exhibit significant diel and seasonal variations. Phytoplankton and zooplankton populations can fluctuate dramatically with tides, light availability, and nutrient inputs. Such dynamic conditions create a complex tapestry of interactions, underpinning the productivity and biodiversity associated with coastal regions. Coastal blooms, while essential for local ecosystems, can sometimes lead to hypoxia or 'dead zones' when decomposing organic matter depletes oxygen levels, illustrating the delicate balance within these environments.

The open ocean, or pelagic zone, with more than 1300 million km^3 and a maximum depth of 11 km, covers the majority of Earth's surface and presents a vast, relatively homogeneous environment where plankton communities exhibit unique characteristics based on depth and latitude. In the upper layers of the open ocean, known as the epipelagic zone, phytoplankton like *Trichodesmium* (together with *Prochlorococcus* and *Synechococcus*) play dominant roles. *Trichodesmium*, a nitrogen-fixing cyanobacterium, thrives in oligotrophic (nutrient-poor) tropical and subtropical waters, forming visible patches that resemble sawdust on the water's surface. These blooms contribute significantly to nitrogen cycling, supporting primary productivity in otherwise nutrient-deficient regions.

From Surface to Depth

Plankton communities also exhibit remarkable vertical stratification, with distinct changes in composition and abundance from the ocean surface to its deepest reaches. This vertical distribution is influenced by factors such as light availability, pressure, and nutrient gradients.

Plankton at the very surface of the ocean (neuston) are adapted to endure high UV irradiance and are usually shielded by photoprotective pigments (Fig. 10.3). In the euphotic zone (from surface to about 200 m, Fig. 10.4), sunlight supports photosynthesis, making it the primary habitat for phytoplankton. Diatoms, dinoflagellates, and cyanobacteria flourish here, driving primary production and forming the base of the marine food web. Below the euphotic zone lies the mesopelagic zone (Fig. 10.4), often called the "twilight zone." Light diminishes rapidly with depth, and phytoplankton are replaced by heterotrophic organisms. Zooplankton, such as mysidacea and gelatinous forms like ctenophores (comb jellies, Fig. 10.5), together with some mesopelagic copepods such as *Pleuromamma* spp., dominate this layer. Many of these species undertake diel vertical migrations, ascending to the euphotic zone at night to feed and descending during the day to avoid predators.

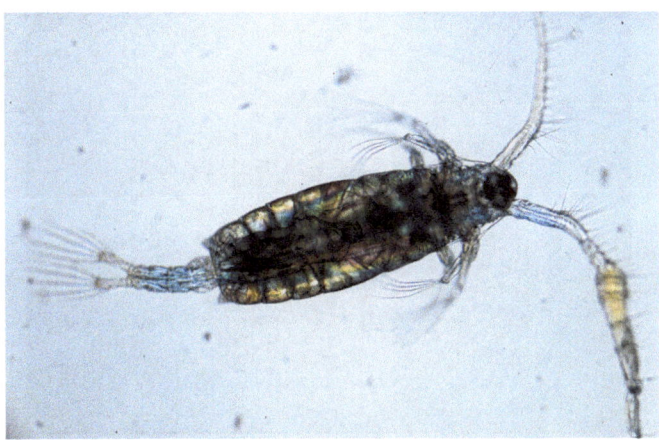

Fig. 10.3 Copepod from surficial waters showing blue photoprotective pigments in the abdomen and antenna. *Labidocera wollastoni.* © Albert Calbet

Fig. 10.4 The different zones of the ocean according to their depths. © Drawings by Albert Calbet

Fig. 10.5 Comb jelly. *Mnemiopsis leidyi* © Albert Calbet

As we venture deeper into the bathypelagic and abyssopelagic zones, light disappears entirely, and pressure increases. The scarce plankton here are adapted to extreme conditions. Organisms like deep-sea jellyfish rely on the few prey available and the detritus falling from above (marine snow) for sustenance. The deep-sea plankton community also includes gelatinous zooplankton like siphonophores, which can reach extraordinary lengths and exhibit intricate colonial structures. The absence of light means that bioluminescence becomes a crucial adaptation for communication, camouflage, and predation.

The benthic boundary layer, where the ocean meets the seafloor, represents another distinct habitat. Here, benthic plankton, including various larval stages of invertebrates and resuspended microorganisms, interact with the sediment. This layer plays a crucial role in nutrient cycling and supports unique communities of deep-sea organisms.

The Benthic-Pelagic Coupling

The true nature of the marine ecosystem is defined by a continuous exchange of matter, energy, and organisms between the benthic and the pelagic zones. This interaction, referred to as benthic-pelagic coupling, is essential for maintaining ecological balance and sustaining the productivity of marine environments. For instance, the larval stages of many reef fish and invertebrates (Fig. 10.6) spend part of their lives as plankton. Also, the symbiotic relationship between corals and microalgae (zooxanthellae) is a prime example, where corals provide habitat and nutrients from waste to the algae, which in turn, through photosynthesis, supply the coral with essential organic compounds. Disruptions, such as coral bleaching due to ocean warming, underscore the fragile link between these zones, where pelagic changes directly affect benthic communities.

Filter-feeding organisms, such as mussels and clams, play a key role in marine ecosystems by filtering water and depositing organic material, which helps recycle nutrients. This process connects the benthic and pelagic zones by nourishing benthic organisms and regulating nutrient levels in the water. In areas with abundant filter feeders, these organisms can prevent harmful plankton blooms and eutrophication. However, when filter-feeder populations are too high, such as in aquaculture facilities, the excess detritus they produce can lead to severe eutrophication and oxygen depletion, disrupting the ecosystem.

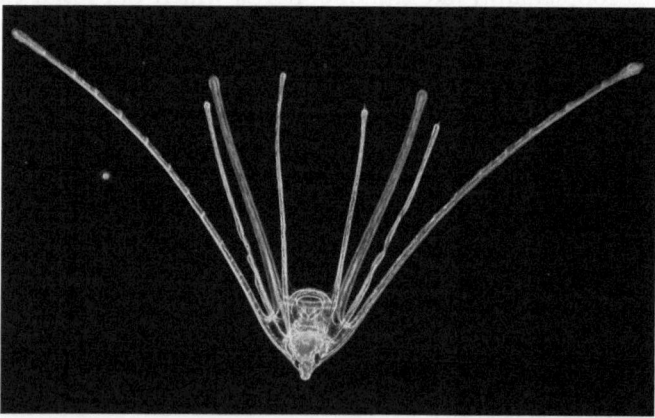

Fig. 10.6 The ophiopluteus larvae of a brittle star belong to a group of organisms that have a phase of their lifecycle in the plankton, known as meroplankton. This is in contrast to holoplankton, which spend their entire life in the plankton. © Albert Calbet

As previously mentioned, marine snow, together with some planktonic organisms, sustains benthic communities, especially in deep-sea areas where sunlight does not reach. Resuspension of these nutrients due to physical forces like currents brings them back into the water column, ensuring the availability of nutrients for pelagic organisms. Additionally, bioturbation—sediment mixing by benthic creatures—promotes nutrient cycling, supporting the productivity of both zones.

Fish, particularly species that feed on benthic organisms and excrete in the pelagic zone, also play a role in nutrient cycling. These fish influence the distribution of organic material, which in turn affects benthic and pelagic ecosystems. Their movement highlights the interconnectedness of marine life.

In conclusion, benthic-pelagic coupling is vital for sustaining ocean productivity, recycling nutrients, and maintaining biodiversity. The health of these zones is interdependent, and understanding their connections is key to protecting marine ecosystems and ensuring the resilience of ocean life.

11

Adrift but not Lost: Dispersal and Colonization Strategies of Plankton

Have you ever wondered how plankton, with their limited motility, can colonize new habitats? The truth is that plankton have developed a remarkable array of dispersal and colonization mechanisms that allow them to exploit distant resources, establish themselves in new habitats, and play a critical role in the health of aquatic ecosystems.

One defining feature of plankton dispersal is their reliance on large-scale ocean currents. Many phytoplankton and zooplankton lack the ability to swim against strong currents. Instead, they depend on a combination of ocean currents, wind-driven circulation patterns, and turbulent mixing to be carried vast distances. This passive drift allows them to exploit favorable conditions for growth and reproduction in different parts of the ocean. However, plankton are not entirely at the mercy of the currents. Some zooplankton species, like copepods, exhibit a fascinating behavior called diel vertical migration. These tiny creatures spend their days in the cool, dark depths of the ocean to avoid visual predators. But as night falls, they embark on a journey upwards, toward the surface waters where phytoplankton blooms often occur. In the process, they become passively dispersed across horizontal currents as they travel vertically through the water column.

© The Author(s), under exclusive license to Springer Nature Switzerland AG 2024
A. Calbet, *Plankton in a Changing World*,
https://doi.org/10.1007/978-3-031-76121-8_11

Even small plankton, with limited swimming ability, can influence their own dispersal to some extent. Certain microplankters possess flagella, tiny whip-like structures that allow them to move slowly in response to light or nutrient gradients. Others benefit from flotation mechanisms that allow them to position themselves strategically for optimal growth. By adjusting their vertical position in the water column, they can be potentially dispersed by currents.

Some plankton species have adopted ingenious strategies to hitch a ride on larger organisms. For instance, barnacles (Fig. 11.1), which possess a planktonic larval stage, can often be found clinging to the bodies of whales and other large creatures. This "hitchhiking" behavior allows them to travel long distances and access new habitats they could not reach on their own. Another fascinating strategy for dispersal involves a journey through the digestive system of a fish. Some zooplankton eggs have a tough outer coating that allows them to survive the harsh environment within a fish's gut. These "internal dispersers" are then deposited in new locations with the fish's waste, effectively colonizing new habitats through a rather unconventional method.

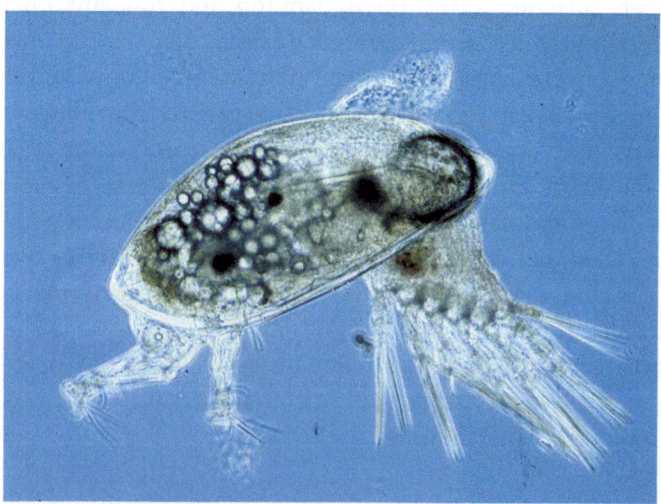

Fig. 11.1   Barnacle larva. © Albert Calbet

Unfortunately, human activities have introduced a new dimension to plankton dispersal with potentially harmful consequences. The discharge of ship ballast water, used to stabilize ships during travel, is a major culprit. Ballast water often contains a cocktail of organisms from the port of origin, including plankton species that may not be native to the destination port. These introduced plankton can disrupt the delicate balance of existing ecosystems, potentially outcompeting native species or introducing diseases. Plastics and other floating human-derived structures can also be vectors for plankton dispersal (Fig. 11.2). Tiny organisms, such as dinoflagellate cysts, can attach themselves to the surface of plastics, hitching a ride across vast distances and potentially introducing invasive species to new environments. The long-term ecological effects of this phenomenon are still being studied, but it highlights the unintended consequences humans can have on the delicate balance of marine ecosystems.

Once plankton reach a potentially suitable habitat, their remarkable ability to colonize comes into play. Plankton are often characterized by high reproductive rates, allowing them to establish themselves in areas

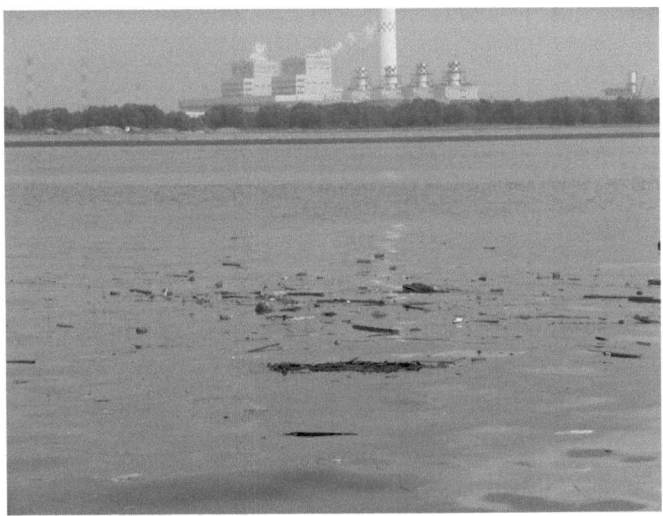

Fig. 11.2 Plastics and other floating substrates are a new way of transport for many organisms of plankton. Singapore waters. © Albert Calbet

with favorable conditions. Some phytoplankton species, for example, can reproduce asexually every few hours, leading to explosive population growth when conditions are right. Zooplankton, in turn, can release large numbers of eggs, ensuring a constant stream of potential colonizers.

Understanding the dispersal patterns of plankton is becoming increasingly important in the face of a changing climate. As ocean temperatures rise and currents shift, the distribution of suitable habitats for plankton species will undoubtedly change. Studying their dispersal mechanisms can help us predict how plankton communities will respond to these changes and how this might affect marine ecosystems.

12

Sailing Through Chaos: Turbulence's Impacts on Plankton

If you are one of the many (like unfortunately myself) that suffer from seasickness, you may find yourself looking horrified the dark clouds preceding a storm, particularly if you have to get into a boat. But for those creatures inhabiting below ocean surface, water motion is an everyday routine. These microscopic drifters have their existence intricately intertwined with the ever-changing nature of ocean turbulence, creating a fascinating interplay of challenges, opportunities, and disadvantages. Before deepening into the interactions of plankton and turbulence let me briefly describe the various manifestations of turbulence encountered in the ocean.

Ocean in Motion: Understanding the Different Scales of Turbulence

Beyond the general understanding of turbulence as a chaotic movement of water, it is important to differentiate between its various scales within the ocean, as they impact plankton differently:

© The Author(s), under exclusive license to Springer Nature Switzerland AG 2024
A. Calbet, *Plankton in a Changing World*,
https://doi.org/10.1007/978-3-031-76121-8_12

1. *Currents* are the large-scale flows of water in the ocean, ranging from kilometers to thousands of kilometers wide and extending for hundreds or even thousands of meters deep. Imagine rivers flowing within the ocean. Currents are primarily driven by large-scale forces like surface winds that drag on the ocean surface, by density differences driven by differences in temperature (warm water is less dense) and salinity (saltier water is denser) that can create pressure gradients that drive currents (these are known as thermohaline circulation) or by the Earth's Rotation that deflects currents due to the Coriolis effect, leading to phenomena like gyres (large circular currents). Currents can transport plankton over vast distances, influencing their distribution patterns. Strong currents can also disrupt feeding and vertical movement for some plankton species.

2. *Large-Scale Turbulence* encompasses eddies and other swirling motions that range from tens to hundreds of kilometers in size. Think of these as giant whirlpools within the ocean. They often arise from instabilities within currents themselves. When currents encounter changes in depth, coastlines, or interact with other currents, their smooth flow can break down into eddies. Large-scale turbulence can be a mixed bag for plankton. It can disrupt their position and increase energy expenditure, similar to small-scale turbulence. However, it can also enhance encounter rates between phytoplankton rich areas and zooplankton, boosting feeding efficiency.

3. *Small-scale turbulence* refers to the chaotic, short-lived motions on the scale of millimeters to meters. This turbulence is generated by a variety of factors like wind stirring the surface water, breaking waves, and interactions between different water layers with varying densities. Small-scale turbulence can be particularly disruptive for smaller, weaker-swimming plankton. They can get tossed around, hindering feeding and vertical movement. However, for some zooplankton, it can also increase encounter rates with food particles, promoting growth.

In analogy, you may envision the ocean as a vast highway, with strong currents akin to lanes guiding its flow. Within this highway, large potholes represent the presence of large-scale turbulence, while the bumpy texture of the road mirrors small-scale turbulence. Plankton traverse this ever-evolving landscape, encountering diverse challenges and opportunities dictated by the scale of motion they confront.

Dancing the Turbulent Tango

At times, turbulence poses a significant challenge for plankton. Imagine a microscopic world where gentle currents suddenly transform into swirling vortices, tossing and turning these tiny organisms with unpredictable force. For some plankton, particularly weaker swimmers like dinoflagellates (Fig. 12.1), or those heavily depending on still waters to detect prey (such as the copepod *Oithona* spp., Fig. 12.2) this constant jostling can disrupt feeding, hinder their ability to maintain their position in the water column, and even lead to physical damage. Additionally, turbulence can increase the energy expenditure of plankton as they struggle against the chaotic flow, potentially impacting their growth and reproduction. However, this is not the whole story. In this seemingly harsh environment, some plankton species have discovered a remarkable ability to not only survive but also thrive amidst the turbulence. This ability hinges on their size and morphology. Many zooplankton species, like copepods, possess elongated bodies or appendages that act like keels, providing them with stability and allowing them to navigate the turbulent currents more effectively. Moreover, some phytoplankters, such as

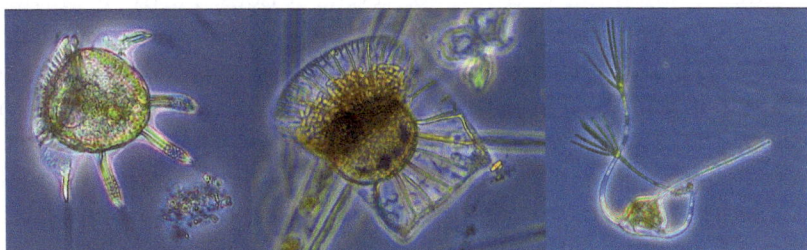

Fig. 12.1 Marine dinoflagellates are usually negatively affected by turbulence. From left to right: *Ceratocorys* sp., *Ornithocercus* sp., *Tripos ranipes* © Albert Calbet

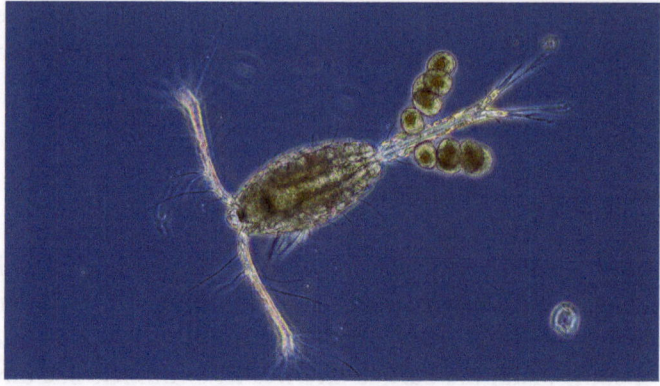

Fig. 12.2 The copepod *Oithona davisae* is highly sensitive to turbulence. ©
Albert Calbet

diatoms, use turbulence to stay the longest in the upper layers of the
water column and to encounter micro-patches of nutrients.

As previously mentioned, one surprising benefit of turbulence for
some plankton lies in its ability to enhance encounter rates between predators and prey. The chaotic mixing brings food closer to zooplankton
increasing their feeding efficiency and potentially leading to higher zooplankton growth rates. This advantage is nullified when turbulence
reaches excessive levels, disrupting feeding currents or rendering mechan-
ical or chemical prey detection impossible. Furthermore, recent research
suggests that some plankton species can non-intentionally "ride" the turbulent currents to their advantage (or detriment).

Understanding the complex relationship between turbulence and
plankton distribution is crucial for several reasons. Firstly, it provides
valuable insights into the health and functioning of marine ecosystems.
By studying how different plankton species, like mesozooplankton and
protists, respond to and interact with turbulence of varying scales, scientists can predict the potential impacts of changes in ocean circulation
patterns or the intensity of turbulence caused by climate change on these
ecosystems. Secondly, studying the potential benefits of using turbulence
for plankton may have applications in aquaculture and conservation
efforts. By understanding how turbulence influences the movement of
plankton larvae, scientists can develop strategies to improve the survival

rates of cultured organisms or facilitate the natural dispersal of endangered species. In conclusion, the dance between plankton and turbulence is far from a simple struggle for survival. It is a complex interplay of challenges and opportunities, where these tiny organisms have evolved remarkable adaptations to not only endure but also potentially exploit the chaotic nature of their environment.

Unveiling the Dance: Tools for Understanding

Scientists utilize a diverse array of tools and techniques to understand the intricate relationship between small-scale turbulence and plankton in the vast ocean.

1. In situ *measurements* in which scientists deploy submersible instruments equipped with high-resolution sensors to directly measure physical parameters like velocity, shear, temperature, and conductivity at specific locations. This allows them to capture the spatial and temporal variations of small-scale turbulence and relate these fluctuations to plankton distribution and behavior. Furthermore, underwater video cameras and other advanced imaging systems are used to visualize plankton abundance and movement patterns within the turbulent flow. This allows researchers to observe how different species react to the chaotic water movement and assess their ability to maintain position or actively navigate the eddies.

2. *Laboratory experiments* under controlled conditions allow scientists mimicking the characteristics of ocean turbulence using specialized equipment. This permits them (us) to isolate the effects of small-scale turbulence from other environmental factors and study the specific responses of different plankton species. Scientists make use also of modern software and technologies for particle tracking. By tagging tiny particles with fluorescent dyes or other tracking mechanisms, they can track their movement within simulated turbulent flow. This helps them understand how turbulence influences the encounter rates between food particles and filter-feeding zooplankton.

3. *Computational modeling* allows for numerical simulations, which combined with powerful computers can be used to create sophisticated models that simulate the dynamics of small-scale turbulence and its interaction with plankton populations. These models incorporate data from field observations and laboratory experiments to predict plankton distribution and behavior under different turbulence regimes.

By combining these diverse approaches, scientists gain a comprehensive understanding of how small-scale turbulence shapes the lives of these microscopic marine organisms. This knowledge not only contributes to a deeper understanding of marine ecosystems but also has potential applications in areas such as aquaculture, fisheries management, and conservation efforts.

13

Plankton Sentinels: Resting Cysts and Dormancy

In summer 2023, we were astonished by the news of worms revived after 46,000 years frozen in Siberian permafrost. While one might assume plankton cannot compete with such longevity records in resting stages, reality surpasses expectations. As they drift through a world of constant change, plankton endure fluctuations in nutrients, temperature, and threats from. In the face of this adversity, remarkable survival strategies emerge, including dormancy and the formation of resting cysts.

Bacteria: Masters of Adaptation

Thriving for billions of years, bacteria have colonized every conceivable niche on Earth, including the extreme environments of the deep sea. Unlike some phytoplankton and zooplankton that form specialized resting cysts, most bacteria do not use this specific strategy. However, they do not simply succumb to harsh conditions. Bacteria possess an arsenal of survival tactics. When faced with limited nutrients, extreme temperatures, or other stressors, some enter a dormancy-like state. Their metabolic processes slow down significantly, but they remain alive and potentially resume growth when conditions improve. Unlike true cysts, they do not develop a thickened cell wall but simply become metabolically inactive. However, some marine bacterial species form cysts, and others, particularly those belonging to the Bacillota phylum, can form endospores. These are highly resistant dormant stages with a thick outer coat, allowing them to withstand harsh conditions during extremely long periods. Perhaps, the record corresponds to some bacteria from deep sediments of the South Pacific, resuscitated after 100 million years. Surprisingly, most of the revived bacteria breathed oxygen, even a light-harvesting bacteria was found in one sample. These bacteria, named *Chroococcidiopsis*, are so incredibly resistant that some suggested their use for terraforming Mars.

While we understand some bacterial dormancy strategies, much remains to be discovered. Researchers are actively investigating the mechanisms bacteria use to enter and exit dormancy. Additionally, the full extent of diversity in dormancy strategies across different bacterial groups needs exploration.

Phytoplankton: The Power of the Cyst

Phytoplankton encompass a diverse range of microscopic algae. Some species, like dinoflagellates and diatoms among others, can transform themselves into a dormant stage when environmental conditions become unfavorable. This transformation involves forming a tough, protective shell or cyst around the organism's cell. Within this cyst (Fig. 13.1), the phytoplankton's life processes slow down dramatically, allowing them to enter a state of suspended animation. This adaptation shields them from harsh realities, like nutrient scarcity, extreme temperatures, or intense sunlight, and also to climatic drastic events associated with global warming. These cysts are believed to be especially important for initiating harmful algal blooms. When exposed to specific light regimes and nutrient levels, cysts emerge, potentially leading to bloom events. The oldest cysts perhaps correspond to a dinoflagellate, *Pentapharsodinium dalei*, that was viable after ca. 100 years in the sediments of the Koljö Fjord (Sweden).

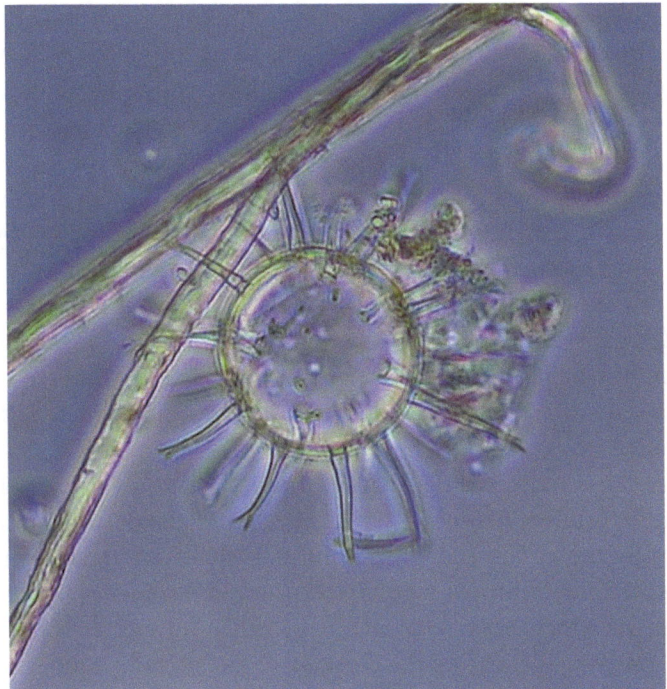

Fig. 13.1 Dinoflagellate empty cyst. © Albert Calbet

Zooplankton: Resting Eggs and Ephippia

Zooplankton are not exempt from the challenges of a fluctuating environment. While some reproduce rapidly to maintain population levels, others, like rotifers, copepods, and cladocerans, adopt a similar strategy to phytoplankton-forming resting eggs or diapausing stages. These dormant stages allow zooplankton to survive periods of adversity, such as food scarcity or unfavorable temperatures. They act as silent sentinels, lying dormant in the sediments or water column, waiting for the opportune moment to reawaken.

Rotifers: Many rotifer species produce resting eggs enclosed in a tough chitinous shell. These eggs can withstand harsh conditions like drought, freezing temperatures, and even complete desiccation. When favorable conditions return, the eggs hatch, releasing new rotifers into the water column. But their story does not end here; some rotifers can dry out their entire body and induce a dormancy state that endures harsh conditions, including complete absence of water.

Copepods: When facing unfavorable conditions, they produce resting or diapause eggs that sink to the bottom. Encased in a protective shell, these eggs can remain dormant for years, triggered to hatch by rising temperatures or specific light regimes. In many species, this is vital to guarantee the natural succession of species. Some other copepod species, particularly large ones from polar waters, such as the genus *Calanus*, do not complete their entire lifecycle during summer. Instead, they enter a dormant state (diapause) in the deep glacial waters until the next summer approaches.

Cladocerans: These peculiar zooplankton, commonly known as water fleas, utilize a unique form of dormancy. Some cladoceran species produce resting eggs encased in a chitinous shell called an ephippium (Fig. 13.2). These ephippia are often adorned with spines or hooks, allowing them to attach to aquatic plants or other submerged objects. The attachment strategy ensures the survival of the eggs during harsh winter months or periods of drought, with hatching occurring when conditions improve. The survival time can be very long; researchers successfully hatched *Daphnia pulicaria* (a freshwater cladoceran) from a resting egg bank hidden in South Central Lake (USA) sediments for over 600 years! The eggs, remarkably, date back to 1418 AD, making them the oldest successfully revived crustacean eggs ever discovered! Imagine, the cladocerans producing those eggs were born years before America was even discovered by Christopher Columbus.

A Window into the Past: Environmental Indicators

Resting cysts act as a historical record within the sediments. Their presence and abundance can reveal valuable information about past environmental conditions. By analyzing the types and quantities of cysts present

Fig. 13.2 *Daphnia pulex*, a freshwater cladoceran, carrying ephippia. ©
Albert Calbet

in sediment cores, scientists can reconstruct past climate patterns, nutrient levels, and even salinity regimes. This information is crucial for
understanding how aquatic ecosystems have responded to past environmental changes and can help us predict how they might respond to future
changes associated with climate change.

  In conclusion, the diverse dormancy strategies employed by plankton
paint a remarkable picture of resilience and adaptation in the microscopic

world. These strategies ensure the survival of these essential organisms in a constantly changing environment and play a crucial role in maintaining the health and diversity of our aquatic ecosystems. By continuing to unravel the mysteries of dormancy in plankton, we gain a deeper appreciation for the intricate web of life that sustains our planet's ocean and freshwater systems.

14

The Paradox of the Plankton: The Kill the Winner Hypothesis

For decades, scientists were perplexed by the paradox of the plankton. Thousands of plankton species, despite having similar needs and competing for the same resources, coexist in the ocean. Traditional theories, like the competitive exclusion principle, suggested only one species should dominate, driving all others to extinction. However, the "Kill the Winner" hypothesis was proposed. The hypothesis suggests that the most dominant species in a microbial community are often targeted by specific predators, viruses, or parasites. Essentially, when a species becomes too abundant, it becomes more vulnerable to predation or parasitism, which helps to maintain biodiversity within the ecosystem.

In short, the mechanism would be the following: The "winners" are fast-growing and efficient, but with vulnerabilities exposed by their very success. Viruses and other predators exploit these weaknesses. As "winners" grow in number, their encounters with viruses increase, leading to higher infection rates and population control. This allows the slower-growing "defenders," who invest in stronger defenses like thicker cell walls or resistance mechanisms, to flourish in the space and resources vacated by the virus-controlled "winners" (Fig. 14.1). The mechanism creates a dynamic equilibrium where no single species dominates,

A. Calbet, *Plankton in a Changing World*,
https://doi.org/10.1007/978-3-031-76121-8_14

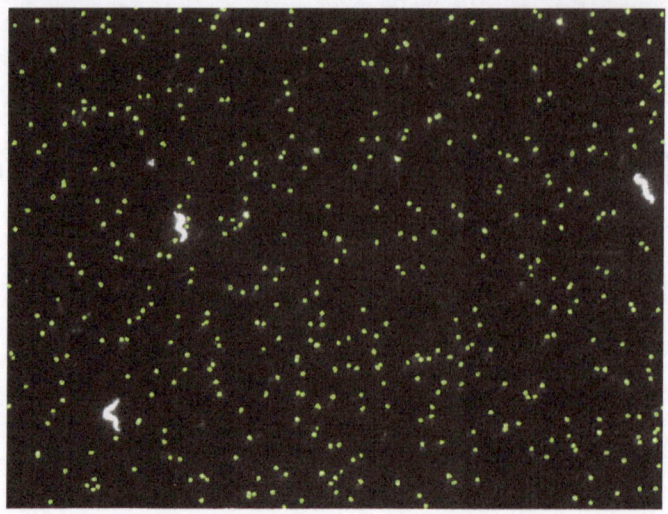

Fig. 14.1 Epifluorescence microscopy of marine viruses. © D. Vaqué and Y. Castillo

fostering the incredible diversity observed in the plankton community. Moreover, by keeping dominant plankton species in check, predation or parasitism helps to promote species diversity, preventing any single species from overwhelming the ecosystem. This, in turn, contributes to ecosystem stability and resilience, ensuring that disruptions to one species do not have catastrophic effects on the entire community. Additionally, the regulation of planktonic populations through the Kill the Winner mechanism optimizes nutrient cycling processes, thereby enhancing ocean productivity and supporting diverse marine life. Furthermore, constant predation pressure can drive evolutionary adaptations in planktonic species, leading to the emergence of traits that enhance their survival and reduce susceptibility to predation or parasitism over time.

The Kill the Winner hypothesis is not just relevant to the ocean depths. It sheds light on similar dynamics in other ecosystems, including the human gut microbiome. While the gut environment offers physical protection from viruses, the principles of "winners" and "defenders" can still play a role in shaping the diverse microbial communities within us. For example, certain gut bacteria may prioritize rapid growth and energy production, while others invest in resilience and immune function. Understanding these dynamics can help us develop better strategies for maintaining gut health and treating related diseases.

While the Kill the Winner hypothesis offers a compelling explanation, it is not without its complexities. For instance, different types of viruses, with varying lifecycles, target specific "winners" and "defenders" in unique ways. Some viruses burst open their hosts immediately ("lytic" viruses), while others integrate into the host genome and replicate later ("lysogenic" viruses). Regarding other predators, such as protozoans, it may be also a selection toward particular prey, not necessarily the most abundant, contradicting the hypothesis. Additionally, the model considers idealized scenarios and does not account for the full spectrum of environmental factors, such as temperature, light, and nutrient availability, that can influence these interactions.

There are indeed alternative hypotheses: piggyback-the-Winner, which suggests that instead of direct killing, lysogenic viruses "piggyback" on abundant hosts, protecting them from other viruses and contributing to their persistence; the resource partitioning and niche differentiation that emphasizes species occupying distinct niches with specific resource utilization, explaining coexistence without invoking viral control; or the neutral models, proposing that random fluctuations and chance events play a significant role in shaping plankton communities, not necessarily preferential virus-mediated selection.

Perhaps, the most contrasting theory is the one that states that the weaker plankton is more susceptible to grazing, as it occurs in larger animals. This hypothesis, which I guess we could call "Kill the Loser," proposes that grazing by larger organisms like zooplankton preferentially targets weaker and stressed plankton individuals. This could be due to various factors, such as:

- Reduced swimming ability: Stressed plankton may have less energy for movement, making them easier prey.
- Nutrient deficiencies: Deficiencies can weaken cell walls and defenses, making them more vulnerable to physical damage.
- Changes in chemical cues: Stressed plankton might emit different chemical signals that attract grazers.

By selectively removing weaker individuals, grazing can create opportunities for slower-growing but more resilient plankton species to thrive. This mechanism, like the Kill the Winner hypothesis, helps explain the coexistence of diverse plankton communities despite shared resource limitations.

In fact, all hypotheses offer plausible explanations for plankton diversity, highlighting different selection pressures in the ecosystem. They are not necessarily mutually exclusive and might work in conjunction, with viruses and grazers, for instance, impacting plankton communities in complex ways.

Part III

Global Environmental Change

15

Understanding Global Change, Global Warming, and Climate Change: A Comprehensive Guide

Along this book I use terms like global change, global warming, and climate change. These terms are often used interchangeably in public discourse, leading to confusion about their distinct meanings and implications. However, these concepts, while interconnected, describe different aspects of the environmental challenges facing our planet. Understanding these differences is crucial for fostering informed discussions and effective actions to address these pressing issues.

Global change is the broadest of these terms, encompassing a wide array of large-scale transformations affecting the Earth. This concept goes beyond environmental alterations to include economic, social, and cultural shifts that occur on a planetary scale. Examples of global change include deforestation, pollution (Fig. 15.1), urbanization, loss of biodiversity, and changes in water cycles, alongside atmospheric changes. It is a holistic term that captures the cumulative impact of human activities and natural processes on the Earth's systems.

Global warming, a more specific term, refers to the long-term increase in Earth's average surface temperature. This rise in temperature is assumed to be primarily driven by human activities that release greenhouse gases into the atmosphere. The burning of fossil fuels, deforestation, and

© The Author(s), under exclusive license to Springer Nature Switzerland AG 2024
A. Calbet, *Plankton in a Changing World*,
https://doi.org/10.1007/978-3-031-76121-8_15

Fig. 15.1 Plastic and other materials accumulation on a Barcelona (Catalonia, Spain) beach after a storm. © Albert Calbet

industrial processes have significantly increased concentrations of carbon dioxide, methane, and nitrous oxide. These gases trap heat in the Earth's atmosphere, creating a "greenhouse effect" that leads to higher global temperatures. Global warming is thus a subset of global change, focusing specifically on the temperature aspect of the planet's transformation. It is a critical concern because the increase in temperature sets off a chain reaction of other environmental changes and stresses.

Climate change encompasses global warming but also refers to a broader array of changes in the Earth's climate system. These changes include not only rising temperatures but also alterations in weather patterns, the frequency and severity of extreme weather events (Fig. 15.2) like hurricanes, droughts, and floods, and shifts in precipitation and seasons. Climate change captures the full spectrum of changes resulting from global warming, as well as other factors that influence the climate, such as volcanic activity and variations in solar radiation. It highlights the dynamic and complex nature of the Earth's climate, where temperature increases are just one piece of a larger puzzle. Climate change affects ecosystems, biodiversity, human health, agriculture, and water resources, making it a comprehensive term for the numerous impacts of a warming planet.

By recognizing the unique aspects of global change, global warming, and climate change, we can better appreciate the complexity of the environmental challenges we face. This understanding is vital for developing

Fig. 15.2 Heavy storms and extreme weather events are expected to increase in certain regions. Picture from Sant Cugat del Vallès (Catalonia, Spain) © Albert Calbet

comprehensive approaches that address not only the symptoms but also the root causes of these changes. Effective communication and education on these topics can empower individuals and communities to take informed actions toward a sustainable and resilient future.

16

The Impacts of Global Warming on Marine Plankton

As global warming accelerates, its impact on marine ecosystems is becoming increasingly evident (Fig. 16.1). The far-reaching consequences of temperature change on plankton populations are profound, disrupting marine food webs, altering biogeochemical cycles, and ultimately affecting global climate regulation. Understanding these impacts is crucial as they cascade through the marine environment, affecting biodiversity, fisheries, and the health of our ocean.

Phytoplankton and the Temperature Increase

Higher temperatures generally increase phytoplankton's metabolic rates, leading to faster growth and higher photosynthetic activity. However, this metabolic boost comes with a catch: it also accelerates cellular respiration, which can deplete energy reserves more rapidly. In warmer waters, phytoplankton may face a delicate balance between enhanced productivity and the increased energy demands of maintaining cellular functions. This metabolic acceleration can also influence nutrient uptake and

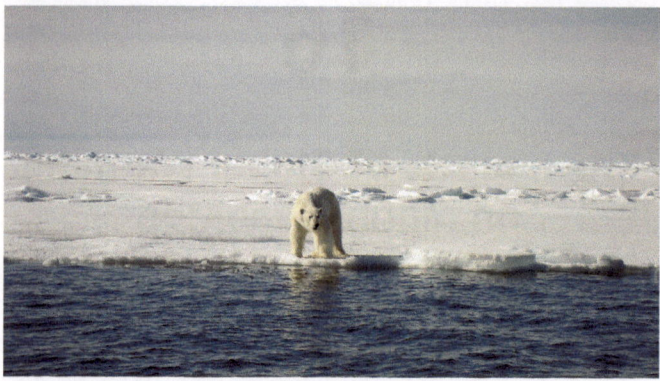

Fig. 16.1 Temperature-driven ice-cap melting will affect the entire ecosystems, from the planktonic inhabitants of marine waters to apex predators, such as polar bears. High Arctic. © Albert Calbet

utilization, shifting the stoichiometric balance of key elements such as carbon, nitrogen, and phosphorus within phytoplankton cells.

The stoichiometric changes induced by temperature rise can have significant ecological repercussions. Phytoplankton typically absorb carbon dioxide (CO_2) during photosynthesis, helping to regulate atmospheric CO_2 levels. However, as temperatures rise, the efficiency of this carbon sequestration can decrease, potentially diminishing the ocean's role as a carbon sink. Furthermore, altered nutrient ratios, usually associated with temperature increases, can affect the nutritional quality of phytoplankton for higher trophic levels, including zooplankton, fish, and ultimately, the entire marine food web. For instance, an imbalance in essential nutrients can lead to poorer growth and reproduction rates in organisms that depend on phytoplankton as their primary food source.

Adaptation to rising temperatures is a critical area of study. Some phytoplankton species exhibit a remarkable capacity to adapt to changing thermal conditions through genetic and phenotypic flexibility. Recent research has shown that certain phytoplankton can adjust their thermal tolerance over relatively short timescales (ca. few months), potentially mitigating some impacts of global warming. These adaptive responses may involve alterations in membrane fluidity, enzyme function, and other physiological traits that enhance survival in warmer waters.

However, the extent and speed of these changes vary widely among different species and populations, suggesting that not all phytoplankton will be equally resilient to temperature increases. Consequently, the overall biodiversity and functionality of marine ecosystems may be at risk if more sensitive species decline while only the most adaptable thrive.

Warmer waters can also alter the stratification of the ocean, creating layers that prevent the mixing of nutrients from deeper waters to the surface where phytoplankton thrive. This nutrient limitation can lead to declines in phytoplankton populations, disrupting the base of the marine food web.

The Domino Effect on Zooplankton

The decline in abundance and stoichiometric changes in phytoplankton significantly affect zooplankton, which are a critical food source for many larger marine creatures, including fish, whales, and seabirds. As phytoplankton availability decreases due to rising temperatures, zooplankton populations also drop. This decline impacts the species that rely on them for sustenance, creating a domino effect that disrupts marine biodiversity and the stability of marine ecosystems.

Temperature also directly influences zooplankton in several critical ways, shaping their physiology, behavior, and overall survival. As water temperatures rise, zooplankton experience an increase in metabolic rates, which accelerates their growth, development, and reproductive cycles, but also their energy demands, which forces zooplankton to consume more food to maintain their increased physiological functions. In warmer conditions, the distribution of zooplankton species often shifts, with some species expanding their ranges while others retreat to cooler, deeper waters. Additionally, warmer waters can affect the buoyancy and swimming behavior of zooplankton, influencing their vertical migration patterns, which are crucial for avoiding predators and accessing food. Heat stress can impair zooplankton immune systems, making them more susceptible to diseases and parasites. The combined effects of increased metabolic demand, altered distribution, disrupted feeding, and heightened vulnerability highlight the complex and often precarious relationship between zooplankton and their changing thermal environment.

Despite these challenges, many zooplankton species exhibit a remarkable capacity to adapt to rising temperatures. Adaptation (in the widest meaning of the word) occurs through genetic changes over multiple generations, whereas phenotypic plasticity (usually called acclimation), where individuals adjust their physiology or behavior, can occur within a single lifetime. For instance, some zooplankton alter their enzyme activities and membrane compositions to function more efficiently at higher temperatures. Behavioral adaptations, such as shifting vertical migration patterns to avoid the warmest surface waters during the day, can also help mitigate thermal stress. Additionally, some species have been observed to evolve shorter lifecycles and faster reproductive rates, allowing them to cope better with fluctuating environmental conditions. However, the capacity for adaptation varies widely among species, with some showing greater resilience than others. The North Atlantic, for example, has seen changes in plankton communities that have altered the distribution and abundance of fish stocks, impacting fisheries and the human communities that depend on them. While these adaptive strategies can enhance the survival of certain zooplankton populations in warmer waters, the overall impact of temperature rise on zooplankton diversity and marine ecosystems remains a critical area of concern. Understanding both the limitations and potential of zooplankton adaptation is essential for predicting the future health of oceanic food webs in the face of climate change.

Shifts in Habitat

Global warming causes planktonic species to shift their habitats. Warmer temperatures drive some species to migrate toward the poles in search of cooler waters, while others may move to deeper layers of the ocean. This displacement can lead to the replacement of native plankton species with those more tolerant of higher temperatures. Such changes can disrupt established food webs, as the new plankton species may not provide the same nutritional value or be as readily consumed by existing zooplankton and fish populations. The replacement of nutrient-rich species with less nutritious ones can lead to lower food quality up the food web, affecting the growth and reproduction of higher trophic levels, including commercially important fish species.

Changes in Seasonal Activities

The temperature increase also affects the timing of seasonal activities of marine plankton, a concept known as phenology. Rising temperatures can shift the timing of phytoplankton blooms, which in turn affects the entire marine food web. For example, if phytoplankton blooms occur earlier in the year due to warmer temperatures, the zooplankton that feed on them may not be synchronized with this shift, leading to a mismatch in food availability. This phenological mismatch can have cascading effects throughout the marine ecosystem, impacting fish that depend on zooplankton and the predators that rely on those fish. Such shifts can alter the seasonal dynamics of marine ecosystems, potentially reducing their resilience and stability.

The Threat of Harmful Algal Blooms

Global warming is contributing to the increasing frequency and intensity of harmful algal blooms (HABs). Certain types of plankton, like dinoflagellates (Fig. 16.2), can proliferate rapidly under warm and nutrient-rich conditions, leading to blooms that produce toxins harmful to marine life and humans. These blooms can cause mass die-offs of fish and other marine animals, contaminate shellfish, and create dead zones where oxygen levels are too low to support most marine life. The Gulf of Mexico, for example, has experienced significant dead zones linked to nutrient runoff and warming waters, highlighting the interconnectedness of human activities, climate change, and marine health.

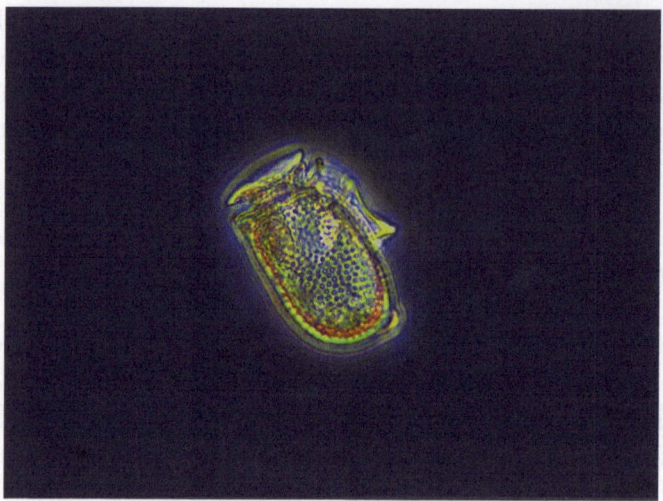

Fig. 16.2 *Dinophysis* sp. is a very toxic dinoflagellate responsible for HABs

The Rise of Jellyfish

Another potential consequence of global warming on marine ecosystems is the rise of jellyfish-dominated environments. Jellyfish thrive in warmer waters and can outcompete fish for food resources, particularly when fish populations are stressed by changes in plankton availability. With fewer predators and competitors, jellyfish populations can explode, leading to blooms that further disrupt marine food webs. These blooms can have severe ecological and economic impacts, including the clogging of fishing nets, the shutdown of coastal power plants, and the decline of fish stocks. The possibility of an ocean dominated by jellyfish (Fig. 16.3) is a stark reminder of the complex and far-reaching effects of climate change on marine ecosystems.

The Importance of Research and Monitoring

Understanding the intricate relationships within the planktonic ecosystem and their response to global warming is crucial. Research and monitoring efforts are essential to track changes in plankton populations and

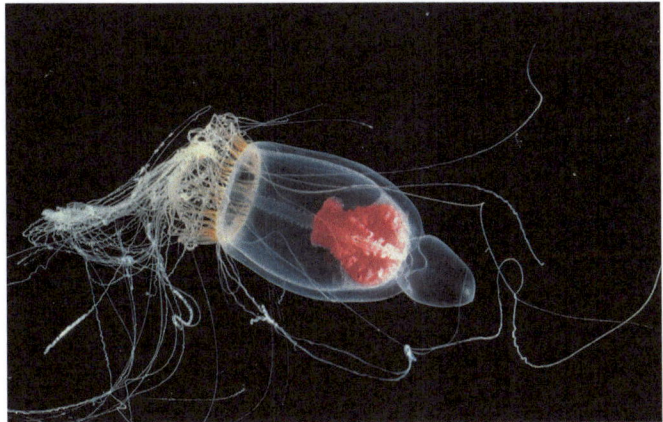

Fig. 16.3   *Leuckartiara octona*. A widespread marine jellyfish. © C. Carré

their effects on the broader marine environment. Technologies such as satellite remote sensing and molecular techniques have advanced our ability to study these tiny but mighty organisms on a global scale, providing valuable insights into their dynamics and helping inform conservation strategies.

17

The Consequences of Smaller Plankton in a Warmer Ocean

In the previous chapter, we discussed the physiological effects of warming on plankton. Here, I will focus on an inevitable and universal effect of temperature: size reduction. The relationship between size and temperature in ectotherms (organisms that do not regulate their body temperature, such as plankton) is called the temperature-size rule (TSR). This rule posits that these organisms tend to grow to smaller sizes when reared in warmer environments. This phenomenon has far-reaching implications for the marine food web and the biological pump, which are critical components of the ocean's ecological and biogeochemical processes. As global temperatures rise due to climate change, understanding the TSR's impact on plankton size is crucial for predicting the future dynamics of marine ecosystems.

Why Plankton Shrink in Warmer Waters?

Several physiological and ecological factors contribute to the size reduction in plankton at higher temperatures. Firstly, higher temperatures accelerate metabolic rates in ectotherms, leading to faster growth but

© The Author(s), under exclusive license to Springer Nature Switzerland AG 2024
A. Calbet, *Plankton in a Changing World*,
https://doi.org/10.1007/978-3-031-76121-8_17

shorter developmental periods. Consequently, planktonic organisms often mature at smaller sizes. Additionally, elevated temperatures can influence the availability and uptake of nutrients. In warmer waters, nutrient recycling processes can become more rapid, but the stratification of the water column may limit nutrient availability in the photic zone, where most plankton reside. This limitation can constrain the growth of plankton, further contributing to smaller sizes, and also favors osmotrophs with higher surface to volume relationships, such as small phytoplankton.

Implications of a Smaller Size

The implications of shrinking plankton sizes are profound for the marine food web. Plankton are primary producers and primary consumers, so any change in their size (Fig. 17.1) affects the entire trophic pyramid. Smaller plankton may alter the feeding efficiency and growth of their predators. For instance, many fish larvae and small marine animals rely on specific sizes of plankton as their primary food source. A shift toward smaller plankton could mean that these predators need to consume more

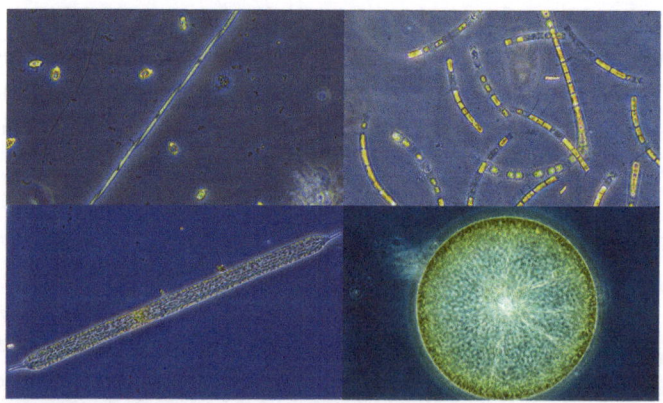

Fig. 17.1 Even within the same group, e.g., the diatoms, we can find representatives of different sizes and shapes. Left up: *Pseudo-nitzschia* sp.; left down: *Proboscia alata*; right up. *Chaetoceros pseudocurvisetus*; right down: *Coscinodiscus* sp. © Albert Calbet

individuals to meet their nutritional needs, potentially leading to higher energy expenditures and altered feeding behaviors. This change can cascade through the food web, affecting higher trophic levels, including commercially important fish species, seabirds, and marine mammals.

Moreover, the shift to smaller plankton can influence the composition and functioning of the zooplankton community. Smaller phytoplankton might favor smaller zooplankton groups, such as microzooplankton, over larger groups like copepods. This shift can impact the efficiency of energy transfer within the food web, as smaller organisms tend to have higher metabolic rates and shorter lifespans, potentially leading to quicker recycling of organic matter within the marine ecosystem. Such change can reduce the overall biomass of zooplankton available for higher trophic levels, thereby diminishing the energy flow to larger marine organisms and affecting the structure and productivity of marine ecosystems, including fish stocks.

The TSR's implications extend beyond immediate food web interactions to biogeochemical cycling, particularly the biological pump (Chap. 18). The biological pump is a crucial process whereby carbon is sequestered in the deep ocean. Phytoplankton absorb carbon dioxide during photosynthesis, and when these organisms die or are consumed, their biomass, including the sequestered carbon, can sink to the ocean depths. Here, it can be stored for long periods, effectively removing carbon from the atmosphere and mitigating climate change.

Smaller plankton sizes due to higher temperatures can disrupt this process. Smaller organisms produce smaller and less dense fecal pellets and organic aggregates, which sink more slowly and are more prone to decomposition before reaching the deep ocean (Fig. 17.2). This reduces the efficiency of the biological pump, as less carbon is transported to and stored in the deep ocean. Additionally, if the plankton community shifts toward species with lower rates of carbon fixation or different sinking dynamics, the overall capacity of the ocean to sequester carbon could be compromised.

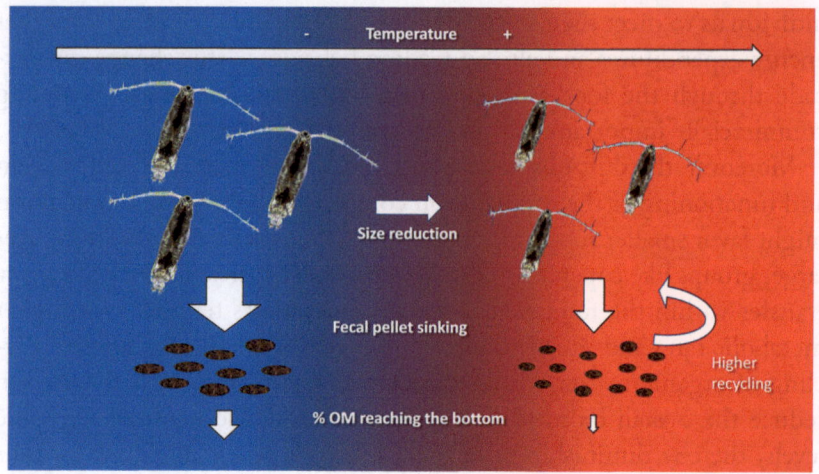

Fig. 17.2 Summary example of the effects of temperature on the biological pump mediated by size reduction. OM = Organic matter. © Albert Calbet

18

The Ocean's Biological Pump: A Crucial Ally Against Global Warming

Far from its peculiar name, the ocean's biological pump is one of the most fascinating and vital processes occurring in our planet's ocean. This natural mechanism plays a significant role in regulating the Earth's climate by sequestering carbon dioxide (CO_2) from the atmosphere and transporting it to the deep ocean. Understanding how the biological pump works and its effectiveness in mitigating global warming is essential for appreciating its contribution to our planet's health.

The Basics of the Biological Pump

At its core, the biological pump (Fig. 18.1) involves the movement of carbon through marine ecosystems. This process can be broadly categorized into two main types: the passive pump and the active pump. Both mechanisms work together to sequester carbon from the atmosphere, but they operate through different biological and physical processes.

The passive biological pump (Fig. 18.2) is driven by the natural sinking of organic matter from the surface ocean to the deep sea. This process begins with phytoplankton absorbing CO_2 during photosynthesis and

© The Author(s), under exclusive license to Springer Nature Switzerland AG 2024
A. Calbet, *Plankton in a Changing World*,
https://doi.org/10.1007/978-3-031-76121-8_18

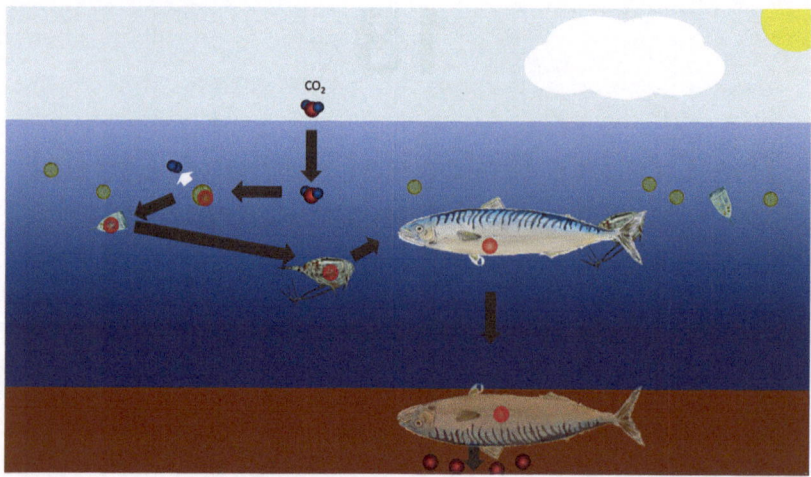

Fig. 18.1 Example of the biological passive pump pathway. The carbon atom is depicted as a red dot (oxygen atoms are blue dots) that enters the food web and ends up buried in the sediment. © Albert Calbet

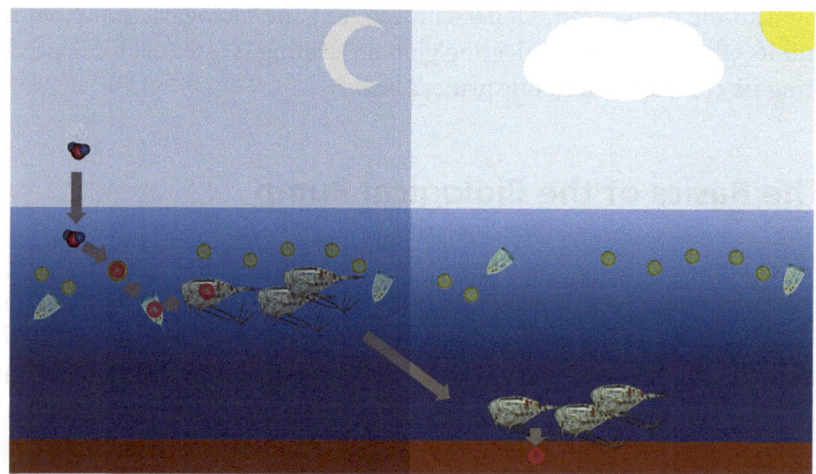

Fig. 18.2 Representation of the active biological pump. At night (left) the zooplankton feeds on phytoplankton containing carbon (red dot) captured from atmospheric CO_2 that entered the ocean. During the day (right), the zooplankton migrates deep into the water column and releases carbon in form of fecal pellets and respiration. © Albert Calbet

converting it into organic carbon as they produce energy and oxygen. When phytoplankton are consumed by zooplankton and other marine animals, the carbon they have stored is transferred up the food web.

A key aspect of the passive pump is the eventual fate of this carbon. When marine organisms die, their remains, along with waste products, begin to sink toward the ocean floor. This sinking organic matter, known as marine snow, carries carbon from the surface to the deep ocean. In deeper waters, the organic matter is decomposed by bacteria, releasing some CO_2 back into the water. However, a portion of this carbon reaches the seafloor, where it can be buried in sediments for hundreds to thousands of years, effectively removing it from the atmospheric cycle.

Although with great variability depending on the region, studies estimate that the passive biological pump transports approximately 5–10 Gt of carbon from the surface to the deep ocean each year. This process is a critical component of the global carbon cycle, helping to regulate atmospheric CO_2 levels and mitigate global warming. Without the passive pump, the concentration of CO_2 in the atmosphere would be significantly higher, exacerbating the greenhouse effect and accelerating climate change.

The active biological pump (Fig. 18.2), on the other hand, is mediated by the vertical migration of zooplankton. Many species of zooplankton undertake daily vertical migrations, moving from the deep ocean to the surface at night to feed on phytoplankton and returning to deeper waters during the day to avoid predators. This migration is one of the largest synchronized movements of biomass on the planet.

During these migrations, zooplankton actively transport carbon to deeper waters. When zooplankton feed on phytoplankton at the surface during the night, they assimilate carbon into their bodies. As they migrate downward during daylight, they respire and excrete waste products, releasing CO_2 and organic carbon at depth. This active transport of carbon to the deep ocean is a vital component of the biological pump.

The efficiency of the active pump is influenced by factors such as the depth and frequency of zooplankton migrations, as well as the metabolic rates of the migrating species. Estimates suggest that the active biological pump can transport an additional 1–2 gigatons of carbon to the deep ocean each year. While this is less than the amount transported by the passive pump, the active pump plays a crucial role in enhancing the overall efficiency of the biological pump and in driving carbon to depths where it is less likely to be quickly returned to the atmosphere.

Impacts of Climate Change

The biological pump is itself not immune to the impacts of climate change. Rising ocean temperatures can stratify the water column, limiting the upward movement of nutrients from the deep ocean to the surface. This nutrient limitation can decrease phytoplankton productivity, reducing the amount of carbon that can be sequestered. Additionally, ocean acidification, driven by increased CO_2 absorption, can affect the growth and calcification of certain phytoplankton and zooplankton species, potentially altering the efficiency of both the passive and active pumps.

To illustrate the significance of the biological pump, consider the carbon sequestration potential of different marine environments. Coastal regions, with their high nutrient input from rivers and upwelling zones, are often hotspots of biological activity and carbon sequestration. In contrast, open ocean regions, especially the oligotrophic gyres with their lower nutrient levels and less biological productivity, largely rely on recycling, and, consequently, present lower carbon sequestration rates. The variability in the biological pump's efficiency across different marine environments highlights the complexity of the ocean's role in the global carbon cycle.

19

Blue Carbon Revolution: Exploiting Plankton to Combat CO_2 Emissions

As our planet copes with the escalating impacts of climate change, the quest for effective solutions has never been more urgent. One promising avenue is gaining attention: Blue Carbon. This concept highlights the pivotal role that oceanic and coastal ecosystems play in capturing and storing atmospheric carbon dioxide (CO_2), thereby mitigating climate change (see previous chapter). By examining the science behind Blue Carbon and understanding its potential, we can appreciate the ocean as more than just a source of beauty and biodiversity—it is also an essential ally in our fight against global warming, or perhaps not?

What Is Blue Carbon?

Blue Carbon refers to the carbon captured by the world's coastal and marine ecosystems, including mangroves (Fig. 19.1), salt marshes, seagrasses, and the vast expanse of the open ocean. These ecosystems are incredibly efficient at sequestering carbon, often outperforming terrestrial forests on a per-area basis. Through photosynthesis, marine plants and planktonic microalgae absorb CO_2 from the atmosphere and convert

Fig. 19.1   Mangrove region in Costa Rica. © Albert Calbet

it into organic matter. When these organisms die or enter the food web and eventually sink out of the photic zone, their carbon-rich residues get buried in sediment, often remaining locked away for centuries or even millennia.

Enhancing Plankton for Carbon Sequestration

Given the effectiveness of natural processes in sequestering carbon, an innovative strategy that has drawn significant interest is the commercial use of plankton to enhance this process. The idea is to help the ocean accumulate more Blue Carbon through artificial means. However, before diving in, it is essential to carefully examine the scientific, environmental, and socio-economic dimensions.

Iron Fertilization: One method to harness the power of plankton is iron fertilization. Researchers discovered that in some large ecosystems, phytoplankton production was limited not by the usual nutrients, nitrogen and phosphorus, but by iron. Therefore, adding iron to the water could produce artificial blooms. Iron acts as a vital nutrient, causing large blooms of phytoplankton that absorb more CO_2. While some early studies have shown promise, there are very serious ecological concerns. Large-scale iron fertilization could disrupt marine ecosystems, leading to harmful algal blooms that deplete oxygen and produce toxins. Such disruptions could affect marine biodiversity and cause the collapse of local fisheries, endangering food security and livelihoods. Additionally, the long-term effectiveness of iron fertilization is still under scrutiny. Some evidence suggests that much of the carbon absorbed by the plankton might not sink deep enough into the ocean to remain sequestered for long periods and could return to the atmosphere, diminishing the strategy's effectiveness. Furthermore, the logistics of distributing iron across vast ocean expanses and monitoring its impacts present significant challenges. There is also the risk of releasing other greenhouse gases like methane during the decomposition of organic material, potentially counteracting the benefits.

Direct Phytoplankton Cultivation: This process offers another interesting approach (Fig. 19.2). This involves growing plankton in controlled environments like bioreactors or designated ocean areas. By optimizing

Fig. 19.2 Scaling marine phytoplankton cultures from laboratory to large bioreactors is a challenge. In the picture we can see different phases of growth of a culture of the marine algae *Rhodomonas salina*. © Albert Calbet

growth conditions, these systems aim to maximize CO_2 absorption. The harvested plankton biomass can be used in various commercial products, such as biofuels or animal feed, or sequestered in deep-sea environments. However, this method is not without its challenges. Maintaining optimal conditions for large-scale plankton growth is technically demanding and energy-intensive, possibly offsetting some environmental benefits. The economic viability of direct cultivation hinges on market demand for derived products and the availability of subsidies or carbon credits. Developing the necessary infrastructure is expensive, and without strong financial incentives, companies might struggle to turn a profit. Ethical considerations also come into play; manipulating natural systems for commercial gain requires a careful balance between innovation and ecological integrity. Ensuring operations do not harm marine ecosystems demands rigorous monitoring and thoughtful planning.

Artificial Upwelling. This is another proposed method to enhance plankton growth for carbon sequestration. This technique involves pumping nutrient-rich deep water to the ocean surface, promoting plankton blooms and boosting primary productivity by mimicking natural upwelling processes. However, the impact of this method on marine ecosystems is complex and not fully understood. While it can increase productivity, it might also alter species distributions and create unforeseen ecological consequences. Energy consumption is also a critical factor for artificial upwelling. If the energy required to pump deep water to the surface comes from fossil fuels, it could negate the carbon sequestration benefits. Additionally, the stability of the carbon sequestered by artificially induced plankton growth remains uncertain.

Socio-Economic Considerations

The commercial use of plankton for carbon sequestration presents both opportunities and challenges from a socio-economic perspective. Developing and deploying technologies for plankton cultivation and sequestration is costly. Economic feasibility depends on market demand, subsidies, and carbon credits, which I particularly think is an aberration. In my opinion, buying "blue" credits to clear the conscience of a polluting company should be absolutely banned. Yet, companies navigate these variables to ensure profitability. Plankton-based industries could stimulate economic growth and create jobs, particularly in coastal communities. However, engaging local stakeholders and ensuring equitable benefit distribution are crucial. Communication and community involvement are key to gaining public support and avoiding conflicts. Moreover, ethical considerations related to manipulating natural systems for commercial gain require a thoughtful and inclusive approach.

To move forward responsibly with the commercial use of plankton for carbon sequestration, it is imperative to adopt the precautionary principle, conduct thorough scientific research, and establish robust regulatory frameworks. Collaboration among scientists, policymakers, industry leaders, and civil society is essential to ensure environmental sustainability, economic viability, and social justice. Proceeding with extreme caution and fostering interdisciplinary dialogue will allow us to harness plankton's potential, contributing meaningfully to the fight against climate change while preserving the health and resilience of our ocean.

20

The Impact of Large-Scale Climatological Events on Plankton Populations

Climatological events like El Niño, La Niña, and the North Atlantic Oscillation (NAO) drive significant changes in oceanographic and atmospheric conditions, profoundly influencing marine ecosystems. Among the various marine organisms, plankton are particularly sensitive to these changes. Understanding how these phenomena affect plankton populations is crucial because plankton form the base of the marine food web and play a significant role in global biogeochemical cycles, including carbon sequestration.

El Niño Southern Oscillation (ENSO) and its Effects on Plankton

The name "El Niño" is Spanish for "the boy" or "the Christ child." It seems that a group of Peruvian fishermen noticing a warm ocean current around Christmas time and deciding to name it after the season's most famous baby. Thanks to El Niño, we get bizarre weather patterns and ocean conditions that throw everything into chaos. It is like Mother

© The Author(s), under exclusive license to Springer Nature Switzerland AG 2024
A. Calbet, *Plankton in a Changing World*,
https://doi.org/10.1007/978-3-031-76121-8_20

Nature's way of saying, "Feliz Navidad (Merry Christmas), everyone! Here's some wacky weather to keep you on your toes!"

El Niño Southern Oscillation (ENSO), jokes aside, is characterized by the periodic warming of sea surface temperatures in the central and eastern Pacific Ocean. During El Niño events, the usual upwelling of cold, nutrient-rich, water along the western coast of South America is suppressed (Fig. 20.1). This upwelling is crucial for the productivity of phytoplankton, which rely on these nutrients for growth. As a result, El Niño typically leads to a marked decline in phytoplankton biomass in this region, which has cascading effects throughout the marine food web, affecting zooplankton and subsequently impacts higher trophic levels, including fish, seabirds, and marine mammals that depend on zooplankton for food.

In contrast, some regions may experience increased productivity during El Niño due to changes in ocean currents and nutrient distribution. For example, the central and western Pacific often see enhanced phytoplankton growth due to altered wind patterns and ocean currents. However, these changes are typically less significant than the nutrient depletions in upwelling regions, leading to an overall decrease in global marine productivity during strong El Niño events. El Niño events influence weather patterns far beyond the Pacific. For instance, during El Niño, phytoplankton productivity increases in the eastern Indian Ocean around Indonesia. Additionally, the Gulf of Mexico and the western

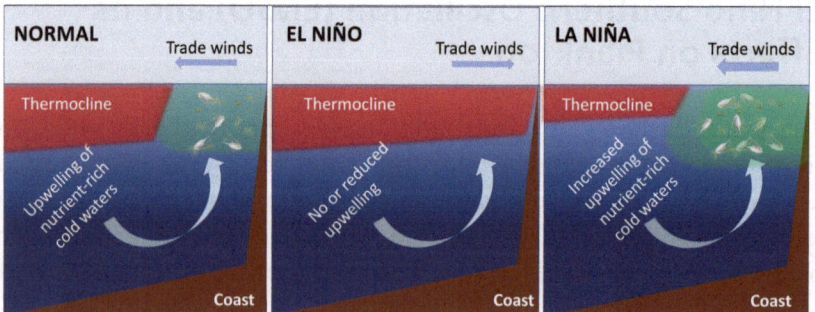

Fig. 20.1 Schematic illustration of normal (left), El Niño (center), and La Niña (right) conditions (right). Note how the upwelling of nutrient-rich waters is prevented during El Niño events and enhanced during La Niña. © Albert Calbet

subtropical Atlantic have seen higher productivity during El Niño events over the past decade. This boost is likely due to increased rainfall and runoff, which bring more nutrients into these regions.

La Niña: The Cooler Counterpart

La Niña (Fig. 20.1), the counterpart to El Niño, involves a cooling of sea surface temperatures in the central and eastern Pacific. This phenomenon generally enhances upwelling and increases the supply of nutrients to surface waters, promoting phytoplankton growth. Consequently, regions like the eastern Pacific experience a surge in phytoplankton productivity during La Niña events. This boost in primary production supports higher populations of zooplankton and other marine organisms, leading to a temporary increase in the abundance of various species within these ecosystems.

However, the impacts of La Niña are not uniformly positive across all regions. In the western Pacific, for instance, the enhanced trade winds associated with La Niña can cause a deepening of the thermocline, potentially reducing nutrient availability in surface waters and negatively affecting phytoplankton growth. Additionally, the increased frequency of storms and changes in precipitation patterns associated with La Niña can influence coastal nutrient inputs and stratification, further complicating the effects on plankton dynamics.

The North Atlantic Oscillation (NAO) and Plankton Dynamics

The North Atlantic Oscillation (NAO) is a climate phenomenon characterized by fluctuations in the atmospheric pressure difference between the Azores High and the Icelandic Low. The NAO influences weather patterns, sea surface temperatures, and ocean circulation in the North Atlantic, which in turn affect plankton populations. During a positive phase of the NAO, stronger westerly winds lead to warmer and wetter conditions in northern Europe and cooler, drier conditions in southern Europe and the Mediterranean. These changes can enhance nutrient mixing in the North Atlantic, leading to increased phytoplankton productivity in some areas.

Conversely, during a negative NAO phase, weaker westerly winds result in colder conditions in northern Europe and milder conditions in the south. This phase can reduce the extent of nutrient mixing, leading to lower phytoplankton productivity in the North Atlantic. Additionally, changes in sea ice cover associated with the NAO can influence light availability for phytoplankton, further affecting their growth. The NAO also impacts the timing and intensity of phytoplankton blooms, which are critical for the lifecycles of many marine species.

A key aspect of NAO's influence is its impact on the Gulf Stream, a powerful Atlantic Ocean current critical to the marine life in the North Sea and that fuels the global circulation pattern, known as the Conveyor Belt (Fig. 20.2). The recruitment success of the copepod *Calanus finmarchicus* (Fig. 20.3), a pivotal zooplankton species, is deeply affected by

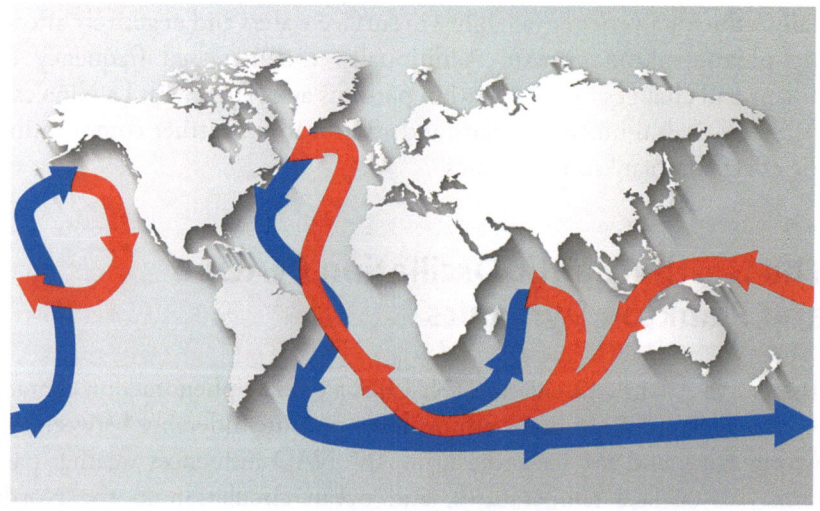

Fig. 20.2 Global circulation pattern or Conveyor Belt. The red arrows indicate the path of warm superficial water, and the blue ones, cold deep water. World map image from Freepik

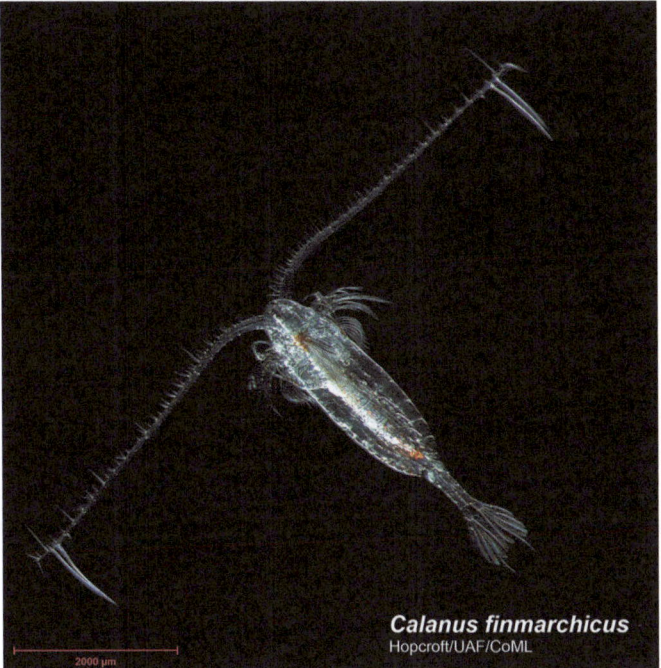

Calanus finmarchicus
Hopcroft/UAF/CoML
2000 µm

Fig. 20.3   *Calanus finmarchicus* © R. Hopcroft

these changes in the Gulf Stream, with cascading effects on the fisheries of cod (Fig. 20.4). *C. finmarchicus*, which overwinters in the North Sea, is crucial for the early life stages of cod, serving as a primary food source. The recruitment of *C. finmarchicus* is tightly linked to sea temperatures and the availability of phytoplankton, both of which are regulated by the Gulf Stream's strength and position.

During a positive NAO phase, the enhanced pressure difference drives stronger westerly winds, intensifying the Gulf Stream's northward flow. This results in warmer sea temperatures in the North Sea, extending the growing season for phytoplankton and thereby potentially boosting the food supply for *C. finmarchicus*. However, this strengthened Gulf Stream can also make it more challenging for young *C. finmarchicus* to migrate southwards from Arctic waters, where they overwinter, limiting their colonization and potentially affecting their population dynamics. These

Fig. 20.4 Cod fish in the marked. © Albert Calbet

fluctuations in *C. finmarchicus* populations directly impact the survival and growth rates of young cod. However, recent studies suggest that the relationship between NAO and *C. finmarchicus* abundance is weakening.

The Pacific Decadal Oscillation (PDO) and Plankton

The Pacific Decadal Oscillation (PDO) is a long-term ocean fluctuation pattern that can significantly impact plankton populations. The PDO operates on a timescale of 20 to 30 years and involves shifts in sea surface temperatures across the Pacific Ocean. During its positive phase, the western Pacific becomes cooler while the eastern Pacific warms, leading to altered ocean currents and wind patterns. These changes can suppress the upwelling of nutrient-rich waters along the western coast of the Americas, similar to El Niño conditions but on a longer timescale. As a result, phytoplankton productivity in these regions can decline, reducing the primary food source for zooplankton and higher trophic levels, including fish and marine mammals.

Conversely, the negative phase of the PDO, which features cooler conditions in the eastern Pacific and warmer conditions in the west, typically enhances upwelling along the American coastlines. This leads to an increase in nutrient availability and a boost in phytoplankton growth. Consequently, zooplankton populations also thrive, supporting a more robust and diverse marine food web. Understanding the PDO's phases helps predict long-term trends in marine productivity and ecosystem health.

The Indian Ocean Dipole (IOD) and Plankton

The Indian Ocean Dipole (IOD) is another major climatological phenomenon that influences plankton populations, particularly in the Indian Ocean. The IOD consists of two phases: positive and negative. During a positive IOD phase, sea surface temperatures in the western Indian Ocean rise, while the eastern Indian Ocean experiences cooler temperatures. This temperature gradient can lead to increased upwelling of nutrient-rich waters in the eastern Indian Ocean, fostering phytoplankton blooms. These blooms provide abundant food for zooplankton and other marine organisms, potentially enhancing fisheries productivity in the region.

On the other hand, the positive IOD can also lead to reduced precipitation and drought conditions in adjacent land areas, affecting coastal nutrient inputs through river discharge. During the negative phase of the IOD, the situation is reversed, with warmer sea surface temperatures in the eastern Indian Ocean and cooler conditions in the west. This can suppress upwelling in the eastern regions, leading to lower phytoplankton productivity and diminished food availability for zooplankton and higher trophic levels. Additionally, increased rainfall during the negative IOD phase can boost nutrient inputs from rivers, temporarily enhancing coastal plankton productivity. The complex interplay between the IOD's phases and regional environmental conditions underscores the importance of monitoring this phenomenon to manage and predict changes in marine ecosystems effectively.

Effects of Climate Change on Climatological Events

Climate change is expected to significantly alter the behavior and impacts of large-scale climatological events. One of the primary concerns is the potential increase in the frequency and intensity of these events due to the warming of ocean temperatures and changes in atmospheric circulation patterns. For instance, warmer sea surface temperatures could amplify the strength of El Niño events, leading to more severe disruptions in global weather patterns, marine ecosystems, and fisheries. Similarly, La Niña events could become more intense, causing more extreme weather conditions, such as increased rainfall and flooding in some regions and severe droughts in others. The PDO, which influences decadal climate variability, may also experience shifts in its phases and duration, affecting long-term marine productivity and ecosystem stability.

Moreover, the IOD could exhibit more pronounced positive and negative phases, altering the distribution of marine resources and affecting the livelihoods of coastal communities. The alteration in the timing, frequency, and strength of these climatological events due to climate change poses significant challenges for predicting and managing their impacts on global and regional scales. Enhanced monitoring and improved climate models are essential to understand these changes better and develop adaptive strategies to mitigate the adverse effects on marine ecosystems, fisheries, and human populations dependent on these resources.

21

Ocean Acidification and Plankton

Since the onset of the Industrial Revolution, the ocean has absorbed approximately one-third of the carbon dioxide (CO_2) emitted by human activities, which helps mitigate atmospheric CO_2 levels. All this tremendous work by the ocean had its consequences. It has led to a steady decline in seawater pH, causing ocean acidification. Ocean acidification begins with the absorption of CO_2 by seawater, forming carbonic acid. This acid dissociates into bicarbonate and hydrogen ions, increasing the water's acidity (lowering its pH). The pH of surface ocean waters has already decreased by about 0.1 units since pre-industrial times, which represents an increase in hydrogen ion concentration by about 26–30%. This seemingly small change represents a significant shift in the delicate balance of ocean chemistry, with profound implications for marine life, including plankton. Future predictions often indicate a significant increase in acidity if current CO_2 emissions continue unabated. Some models do predict that by 2100, the increase in hydrogen ion concentration could reach or exceed 150%, corresponding to a further significant drop in pH levels.

A. Calbet, *Plankton in a Changing World*,
https://doi.org/10.1007/978-3-031-76121-8_21

Phytoplankton: The Base of the Marine Food Web

Phytoplankton are responsible for roughly half of the global photosynthetic activity on Earth, converting CO_2 into organic carbon and releasing oxygen. However, increased acidity can interfere with the photosynthetic machinery of phytoplankton. The pH levels greatly impact photosynthesis by influencing key processes such as enzyme activity, CO_2 availability, chloroplast function, nutrient uptake, and energy production. When pH is optimal, enzymes like RuBisCO work efficiently, and plants can absorb CO_2 effectively. It also helps chloroplasts, the site of photosynthesis, function properly by maintaining the necessary conditions for energy production. Additionally, the right pH ensures essential nutrients are available for the plant. On the other hand, if pH levels are too high or too low, these processes can be disrupted, reducing the plant's ability to perform photosynthesis effectively.

For instance, some species of diatoms, a major group of phytoplankton, show reduced growth rates and altered nutrient uptake under acidic conditions. We should take into account, however, that the deleterious effects on diatoms are evident at very low pH levels, only achieved in the future, under the most pessimistic predictions of high-emissions scenarios. Conversely, some cyanobacteria, which thrive in more acidic environments, may proliferate, potentially leading to harmful algal blooms (HABs). These blooms can produce toxins that devastate marine life and human health, illustrating a direct link between ocean acidification and ecological disruption.

Coccolithophores, a group of calcifying phytoplankton, are particularly sensitive to ocean acidification due to their reliance on calcium carbonate to form their intricate shells, or coccoliths. Increased acidity in seawater reduces the availability of carbonate ions needed for coccolith formation, leading to thinner and more fragile shells. This can impair their ability to survive and thrive, disrupting marine food webs and carbon cycling. However, some studies have indicated that certain strains of coccolithophores may exhibit resistance to acidification, maintaining their calcification rates despite the changing conditions. This variability suggests that while ocean acidification poses a significant threat to coc-colithophores, there is potential for adaptation and resilience within some populations.

Zooplankton Vulnerability to Acidification

Zooplankton, particularly those with calcium carbonate structures, are directly threatened by ocean acidification. Calcifying organisms, such as foraminifera and pteropods, rely on calcium carbonate to build their shells. As seawater becomes more acidic, the availability of carbonate ions necessary for shell formation decreases, causing these organisms to produce thinner, weaker shells. For example, studies have shown that the shells of planktonic snails and pteropods (Fig. 21.1) dissolve when

Fig. 21.1   Pteropod (*Limacina retroversa*) © R. Hopcroft

exposed to pH levels predicted for the end of this century. This vulnerability not only threatens the survival of individual species but also impacts the predators that rely on them, such as salmon and other commercially important fish. Also, some other organisms that rely on calcium carbonate for their skeletal structures, such as echinoderms (Fig. 21.2), will struggle in more acidic water.

Particularly, research conducted in the Southern Ocean has provided stark evidence of the impact of ocean acidification on pteropods. Scientists found that the shells of pteropods (sea butterflies) were already showing signs of dissolution in present-day conditions, indicating that the effects of acidification are already manifesting in natural environments. This has profound implications for the Southern Ocean food web, where pteropods are a crucial food source.

Furthermore, non-calcifying zooplankton are not immune to the effects of acidification. Changes in water chemistry can impair sensory

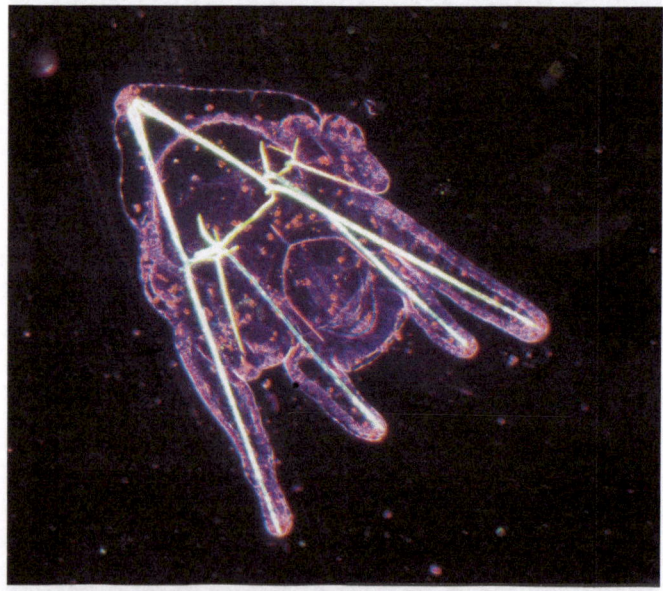

Fig. 21.2 The echinoderm larvae are also very susceptible to acidification. The image shows an ophiuroid larva highlighting the mineral skeleton. Differential interference contrast microscopy. © Albert Calbet

and behavioral functions, making it harder for these organisms to evade predators, find food, and reproduce. For instance, studies on copepods have demonstrated reduced hatching success and altered development rates under strong acidified conditions. These changes, although not evident at present pH levels, can lead, if pH drastically drops, to population declines and disrupt the timing of critical lifecycle events, affecting entire marine ecosystems. Regarding protozoans, evidence suggests that they are more sensitive to alkalinization than to acidification. This sensitivity might serve as a protective mechanism for phytoplankton blooms, which increase the water's pH through photosynthesis. Further research is necessary to explore this phenomenon in detail.

Water acidification also interacts with other global change stressors, such as temperature, oxygen availability, heavy metals, etc., increasing their negative effects. However, there are instances where acidification may mitigate the effects of certain pollutants. For example, a multigenerational study on the copepod *Tigriopus japonicus* found that ocean acidification alleviated nickel toxicity. Under high CO_2 conditions, the copepods exhibited improved survival rates compared to those exposed to nickel alone. This surprising finding suggests that the acidified environment may induce physiological or biochemical changes that enhance the copepod's tolerance to nickel. However, this increased tolerance came at a cost: the copepods displayed a loss of transcriptome plasticity during recovery, indicating a reduced ability to respond to environmental changes. This trade-off highlights the complexity of synergistic effects, where short-term benefits may lead to long-term vulnerabilities.

22

Nutrient Availability and Its Consequences for Plankton in a Changing World

All plants require nutrients to survive and prosper, and phytoplankton are not an exception. Global environmental change is altering nutrient availability in the world's ocean. This shift has significant implications for marine plankton, which rely on a balanced supply of nutrients to sustain their growth and maintain ecosystem health. The changes in nutrient dynamics encompass a range of phenomena, including eutrophication, nutrient depletion, nutrient imbalances, and hypoxia. Understanding these processes and their effects on plankton is crucial for predicting the future of marine ecosystems.

Eutrophication: A Double-Edged Sword

Eutrophication occurs when water bodies receive an excess of nutrients, primarily nitrogen and phosphorus, often from agricultural runoff, wastewater discharge, and industrial activities. This influx of nutrients can lead to the rapid growth of phytoplankton, resulting in algal blooms (Fig. 22.1). While these blooms might initially seem beneficial due to increased primary production, they often lead to detrimental outcomes.

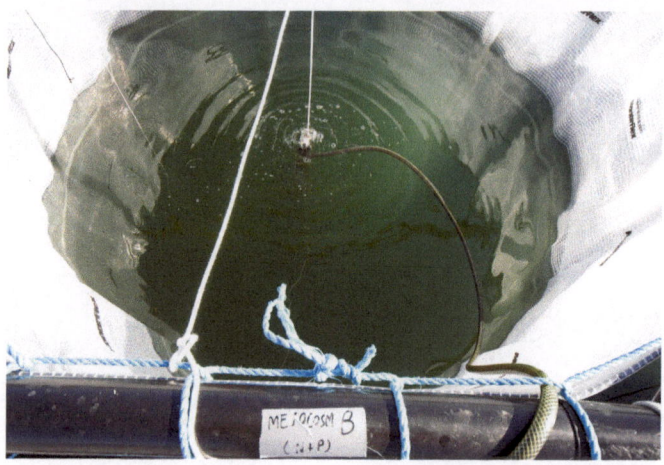

Fig. 22.1 Image of a eutrophic mesocosm after nitrogen and phosphorus addition in an experiment in Bergen (Norway) Mesocosms Facility (MESOAQUA). © Albert Calbet

For instance, the Gulf of Mexico has experienced significant eutrophication due to nutrient runoff from the Mississippi River. The resulting algal blooms have led to hypoxic conditions, creating one of the world's largest "dead zones," where oxygen levels are too low to support most marine life. This has caused substantial declines in fish and shrimp populations, severely impacting local fisheries and biodiversity.

Nutrient Depletion: Starving the Base of the Food Web

Nutrient depletion occurs in areas where nutrient inputs are low or where the nutrient supply is rapidly exhausted by high phytoplankton productivity. In oligotrophic (nutrient-poor) regions of the ocean (Fig. 22.2), phytoplankton growth is limited by the availability of key nutrients such as nitrogen, phosphorus, and iron. Climate change can exacerbate this issue by altering ocean stratification. Warmer surface waters can become more stratified, reducing the mixing of nutrient-rich deep waters to the surface where phytoplankton reside.

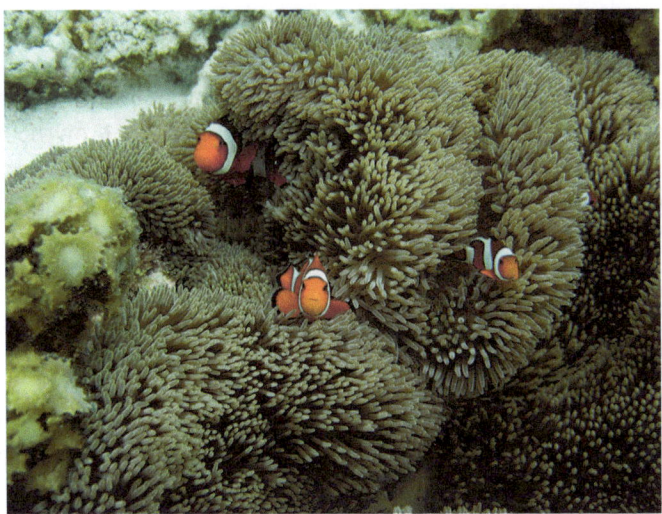

Fig. 22.2   Nutrient-poor waters usually show a higher transparency and diversity. Underwater picture taken during free diving at Komodo coast. © Albert Calbet

For example, in the subtropical gyres of the Pacific and Atlantic Oceans, increased stratification has led to nutrient depletion in surface waters, causing a decline in phytoplankton biomass and primary production. This reduction in primary production has cascading effects on the entire marine food web, from zooplankton to top predators.

Nutrient Imbalances: Disrupting the Redfield Ratio

Nutrient imbalances, particularly deviations from the Redfield ratio, have also effects on plankton communities. The Redfield ratio describes the typical atomic ratio of carbon, nitrogen, and phosphorus (106:16:1) found in marine phytoplankton. When the availability of these nutrients deviates from this ratio, it can lead to suboptimal conditions for phytoplankton growth.

For instance, in the coastal waters of the Baltic Sea, excessive nitrogen inputs have led to phosphorus limitation, resulting in frequent cyanobacterial blooms that can produce harmful toxins. These blooms can disrupt marine ecosystems and pose serious health risks to humans and animals. Conversely, in some regions of the Southern Ocean, iron limitation is a significant constraint on phytoplankton growth. Iron fertilization experiments have shown that adding iron to these waters can stimulate large phytoplankton blooms, potentially enhancing carbon sequestration, though this approach is not without ecological risks.

Hypoxia: A Silent Killer

Hypoxia, often a consequence of eutrophication, represents another significant threat to plankton. When oxygen levels in the water drop below critical thresholds, many aerobic organisms, including zooplankton, fish, and benthic invertebrates, struggle to survive. Hypoxic conditions can lead to the collapse of local plankton populations, especially those that are less tolerant of low oxygen environments.

For example, the Chesapeake Bay has experienced recurring hypoxic events due to nutrient runoff and subsequent eutrophication. These events have led to significant declines in zooplankton populations, disrupting the food web and affecting fish species such as striped bass and blue crabs that rely on zooplankton for food.

Global Primary Production: A Shifting Landscape

Global primary production is also experiencing significant changes due to these factors. Satellite observations and field studies have shown that global marine primary production has been declining in some regions while increasing in others. For instance, studies indicate a decline in primary production in the North Pacific Subtropical Gyre and the tropical Atlantic Ocean, attributed to increased stratification and reduced nutrient upwelling. In contrast, some areas of the Arctic Ocean are seeing increases in primary production due to reduced ice cover, which allows more light to penetrate the water and enhances nutrient availability from melting ice. However, these regional increases are often temporary and may not compensate for declines elsewhere.

The consequences of changing primary production are far-reaching. Reduced primary production in nutrient-depleted regions leads to lower phytoplankton biomass, which affects the entire marine food web. Zooplankton, which feed on phytoplankton, face food shortages, leading to declines in their populations. This, in turn, impacts fish and other marine animals that depend on zooplankton as a food source. For example, declining phytoplankton populations in the North Pacific have been linked to reduced survival rates of young salmon, affecting commercial fisheries and the communities that rely on them.

The interplay between these factors creates a complex and often unpredictable environment for marine plankton. For example, in the North Atlantic, shifts in the distribution of phytoplankton have been observed due to changes in nutrient availability and ocean temperature. Phytoplankton blooms are occurring earlier in the year, affecting the timing of zooplankton reproduction and subsequently the entire food web. In contrast, the localized increase in primary production in the Arctic Ocean is altering the structure of these marine ecosystems, with potential implications for species such as polar bears and seals that rely on fish and other marine organisms for food.

23

Global Change and Salinity: Implications for Plankton

We have all heard the alarming news about melting ice and its impact on rising sea levels, along with the severe socio-economic implications. However, we often overlook that this influx of freshwater not only elevates the ocean's height but also significantly affects its salinity. The decrease in salinity can disrupt marine ecosystems, alter ocean currents, and impact weather patterns globally. Understanding these changes is crucial for preparing for the full spectrum of consequences that come with our planet's rapidly shifting climate.

The Shifting Balance of Salinity

Salinity, or the concentration of salts in seawater, is a critical parameter that influences marine life, including plankton. As global change continues to modify the hydrological cycle, the impacts on ocean salinity and, consequently, on plankton communities will be profound. Bear in mind, for instance, that only the Thwaites Glacier (the "Doomsday Glacier"), the widest glacier on Earth at approximately 120 km (80 miles) wide, in West Antarctica, if it collapses entirely, sea levels could rise by 65 cm (25 inches); if the whole Antarctica's ice melted at once, the sea level would rise 58 m (about 190 feet). Currently, Thwaites contributes about 4% of global sea-level rise, losing around 50 billion tons of ice annually, a rate that has doubled in the past 30 years. All this freshwater getting into the sea must have a significant effect, for sure.

Salinity in the ocean is primarily driven by the balance between freshwater inputs, such as rainfall and river discharge, and the removal of freshwater through evaporation. Climate change is intensifying this cycle. Warmer temperatures increase evaporation rates in some regions, leading to higher salinity levels, while melting ice caps (Fig. 23.1) and glaciers (Fig. 23.2), along with increased precipitation (Fig. 23.3) and river run-off, are freshening other areas of the ocean. These shifts are creating regions with altered salinity patterns, which in turn affect marine ecosystems.

Fig. 23.1 Ice plaques in the high Arctic Ocean during melting season. If the entire Arctic floating sea ice would melt, the sea level would not be significantly affected (like when an ice cube melts in your drink). However, the changes in water salinity would be important. © Albert Calbet

Fig. 23.2 Image of the Grey Glacier in the Torres del Paine, Chile. © Albert Calbet

Fig. 23.3 Water runoff due to heavy rains off Barcelona, Catalonia, Spain. Note the brownish area on the top right of the image. © Albert Calbet

Plankton: Sensitive Indicators of Salinity Changes

Plankton are highly sensitive to changes in salinity. Different species have specific salinity ranges they can tolerate. Euryhaline species can withstand a wide range of salinities (Fig. 23.4), while stenohaline species can only

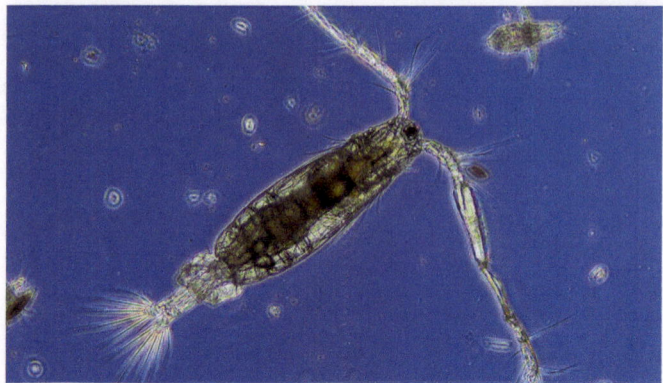

Fig. 23.4 *Paracartia grani* inhabits estuaries and coastal systems; therefore, it has a wider range of salinity tolerance than other more oceanic species. © Albert Calbet

survive within a narrow salinity range. Changes in salinity can thus force shifts in species composition, favoring euryhaline species in fluctuating environments and potentially leading to the decline of stenohaline species.

Phytoplankton are particularly sensitive to salinity variations. For example, studies have shown that increased salinity can inhibit the growth of certain diatoms and green algae while favoring other groups such as cyanobacteria, which are more tolerant of salinity fluctuations. This shift in phytoplankton communities can alter the overall productivity and nutrient cycling within marine ecosystems. In coastal estuaries where freshwater inputs are increasing due to higher rainfall and river flow, phytoplankton communities can shift dramatically. Species that are more tolerant of lower salinity levels may become dominant, altering the overall composition and function of the ecosystem.

Zooplankton: Struggling to Adapt

Zooplankton are affected by changes in salinity as well. Many zooplankton species have limited tolerance to salinity fluctuations, as their osmoregulatory mechanisms—the processes by which they maintain internal salt and water balance—can be easily overwhelmed. For example, cope-pods (Fig. 23.4), a crucial group of zooplankton, show decreased repro-ductive success and survival rates in waters with salinity levels outside their optimal range. This can lead to declines in their populations, which in turn affects the organisms that prey on them, including fish larvae and other marine predators.

Research has shown that salinity changes can cause significant shifts in zooplankton communities. For instance, in the Baltic Sea, a combination of increased river runoff and reduced salinity has led to shifts in zoo-plankton communities, with freshwater-tolerant species becoming more prevalent. This shift has repercussions up the food web, affecting fish spe-cies such as herring and sprat, which rely on specific types of zooplankton as their primary food source.

Regional Variations and Complex Interactions

The effects of salinity changes on plankton are not uniform and can vary greatly depending on the region. Here, I show you some examples.

In the Baltic Sea, a combination of increased river runoff and reduced salinity has led to shifts in phytoplankton communities, with diatoms and dinoflagellates being replaced by cyanobacteria. These cyanobacteria blooms can produce toxins harmful to marine life and human health, illustrating the direct link between salinity changes and ecosystem health. Similarly, in the Gulf of Alaska, freshening of the surface waters due to glacial melt has been associated with changes in the timing and composi-tion of phytoplankton blooms. These shifts can impact the entire marine food web, affecting everything from microscopic zooplankton to large marine mammals that depend on robust fish populations.

In tropical regions like the Great Barrier Reef, changing salinity pat-terns due to increased freshwater runoff from heavy rainfall and tropical storms are influencing the composition of plankton communities. These changes can affect the coral reef ecosystems that rely on specific types of plankton for food. Altered salinity can also stress coral symbionts, like zooxanthellae, which can lead to coral bleaching events. The health of coral reefs is closely linked to the stability of plankton populations, and any disruption can have far-reaching consequences for the biodiversity and tourism industries dependent on these ecosystems.

In the Indian Ocean, variations in salinity are driven by monsoonal cycles and changing precipitation patterns. Studies have shown that these changes influence the distribution and abundance of plankton, with potential impacts on fisheries. For example, during periods of lower salinity, there can be an increase in certain phytoplankton species that are less nutritious for zooplankton, leading to a decline in zooplankton populations and affecting fish stocks that rely on them.

Salinity changes can also interact with other stressors related to global change, such as ocean acidification and warming. Increased salinity combined with higher temperatures or pollutants can exacerbate stress on plankton, making it even more difficult for them to survive and reproduce. Conversely, in regions where salinity is decreasing due to freshwater inputs, the combined effects of lower salinity and ocean acidification can create complex physiological challenges for calcifying plankton species, such as coccolithophores, which rely on carbonate ions to build their shells.

Broader Ecological Consequences

These physiological stresses on plankton can lead to broader ecological consequences. Plankton form the base of the marine food web, and any disruption to their populations can ripple through the ecosystem. Changes in phytoplankton composition and abundance can affect the types and amounts of food available to zooplankton, impacting fish species that rely on zooplankton as a primary food source. Commercial fisheries, which depend on the health and stability of these food webs, may face significant challenges as a result of altered plankton dynamics driven by salinity changes.

The impacts of salinity changes on plankton also have implications for biogeochemical cycles, particularly the carbon cycle. Changes in salinity can affect phytoplankton growth rates and community composition, potentially altering the efficiency of the carbon pump (Chap. 18). For example, if salinity changes lead to a decline in larger phytoplankton species that are more effective at carbon sequestration, this could reduce the ocean's capacity to absorb carbon dioxide, exacerbating global warming.

24

The Underappreciated Threat of Ultraviolet Radiation to Marine Plankton

UV radiation, in moderate intensities, gives us the desired summer tan, albeit not without, many times, first enduring the annoying red and itchy sunburn phase. However, UV, particularly UV-B (280–315 nm), has profound effects on marine ecosystems as well, especially plankton, which are highly sensitive to UV radiation. As climate change continues to affect the ozone layer and water column characteristics, the impacts on UV radiation levels in the ocean, and consequently on plankton communities, will be substantial.

The Changing Nature of UV Radiation

UV radiation is a natural part of sunlight, but its intensity and penetration in the ocean can be significantly influenced by global environmental changes. The depletion of the stratospheric ozone layer, primarily due to chlorofluorocarbons (CFCs) and other ozone-depleting substances, has led to increased levels of UV-B radiation reaching the Earth's surface. Although international agreements like the Montreal Protocol, negotiated in 1987 and further revised in subsequent years, have significantly curbed the emission of these substances, the recovery of the ozone layer is slow and uneven, and elevated UV-B levels persist in many regions. Climate change also plays a role, as greenhouse gas emissions contribute to global warming, which affects stratospheric temperatures and ozone distribution, slowing the ozone layer's recovery.

In addition to ozone depletion, climate change can affect UV radiation penetration in the ocean. For instance, increased stratification of the water column, caused by warming surface temperatures, can reduce the mixing of UV-absorbing substances from deeper waters to the surface. Moreover, changes in the concentration of dissolved organic matter (DOM) and particulate matter, influenced by factors such as land runoff and melting ice, can alter the transparency of seawater, affecting how deeply UV radiation penetrates.

Impacts on Phytoplankton

Phytoplankton are particularly vulnerable to increased UV-B radiation. UV-B can damage cellular components, including DNA, proteins, and lipids, leading to impaired photosynthesis and growth. Studies have shown that exposure to elevated UV-B levels can reduce the photosynthetic efficiency of diatoms and cyanobacteria, two major groups of phytoplankton. This reduction in photosynthesis can lead to lower primary production, which impacts the entire marine food web. Phytoplankton are also critical for carbon cycling, and any decline in their populations can affect the ocean's ability to sequester carbon dioxide, exacerbating global warming.

Specific species of phytoplankton exhibit varying levels of sensitivity to UV-B radiation. For instance, the diatom *Thalassiosira pseudonana* has been observed to experience reduced growth rates and photosynthetic activity under increased UV-B exposure. Similarly, the cyanobacterium *Synechococcus*, an important primary producer in many marine ecosystems, shows significant declines in photosynthetic efficiency and DNA integrity when exposed to high UV-B levels. These disruptions can lead to shifts in community composition, favoring UV-resistant species and potentially altering ecosystem dynamics.

Impacts on Zooplankton

Zooplankton are also affected by UV radiation. Many zooplankton species, such as copepods and krill, rely on behavioral adaptations to avoid UV exposure, such as migrating to deeper waters during daylight hours. However, increased UV-B penetration due to changes in water transparency can overwhelm these avoidance strategies, leading to direct DNA damage, reduced reproductive success, and increased mortality rates. For instance, studies in the Southern Ocean have shown that krill larvae (Fig. 24.1) exposed to elevated UV-B levels exhibit slower development and higher mortality rates, which can impact the populations of species that rely on krill as a food source, including fish, seabirds, and whales.

Fig. 24.1 Krill (*Meganyctiphanes norvegica*). © R. Hopcroft

Shielding the Sun: Neustonic Adaptations to UV Radiation

Surficial plankton, known as neustonic organisms, inhabit the uppermost layer of the ocean, exposed to high levels of sunlight and UV radiation. To protect themselves from the potentially harmful effects of UV radiation, these tiny organisms have evolved a fascinating array of adaptations.

One of the most common adaptations is the production of mycosporine-like amino acids (MAAs). These small molecules act as natural sunscreens, absorbing harmful UV rays and dissipating the energy as harmless heat. MAAs are particularly effective because they can be synthesized or acquired through the food web, allowing a wide range of plankton species to benefit from this protective mechanism. Another intriguing adaptation is the ability to produce protective pigments. Some plankton, like certain species of phytoplankton and zooplankton (Fig. 24.2), can synthesize astaxanthin and carotenoids, which not only give them a distinctive color but also provide UV protection. These pigments absorb and neutralize UV radiation, reducing the risk of cellular damage. Additionally, these pigments can also play a role in photosynthesis, helping plankton convert sunlight into energy more efficiently while simultaneously protecting them from UV-induced harm. Moreover, the extracellular

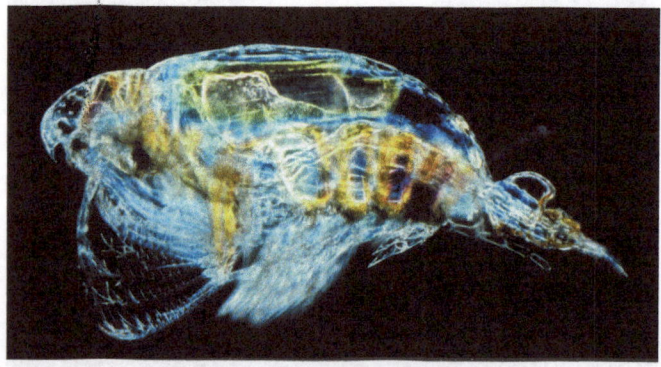

Fig. 24.2 The blue coloration of many neustonic copepods is due to photoprotective pigments. *Labidocera wollastoni.* © Albert Calbet

polymeric substances (EPS) secreted by some plankton offer another layer of defense. These gelatinous substances can form a protective barrier around the cells, reducing UV penetration and preventing direct damage to their DNA and other vital cellular components.

Regional Variations and Complex Interactions

The effects of increased UV radiation on plankton are not uniform and can vary greatly depending on the region and local conditions. In polar regions, the seasonal thinning of the ozone layer, combined with high surface reflectance from ice and snow, can lead to particularly high levels of UV-B radiation during the spring and summer months. This can cause significant stress to phytoplankton and zooplankton communities in these areas, leading to shifts in species composition and productivity. For example, in the Antarctic, increased UV-B levels have been linked to declines in phytoplankton biomass, which affects the entire Southern Ocean ecosystem. In this water UV-B can penetrate up to 50 m depth and its effects on plankton are already apparent at 20–30 m.

In temperate and tropical regions (Fig. 24.3), the interplay between UV radiation and other environmental stressors can create complex challenges for plankton. Increased sea surface temperatures can enhance stratification, limiting the mixing of nutrient-rich deeper waters with the surface and exacerbating the effects of UV radiation on surface-dwelling plankton. Additionally, coastal areas experiencing higher runoff from increased precipitation can see changes in water transparency, affecting UV penetration and its impact on plankton.

Interaction with Other Stressors and Ecological Consequences

Increased UV radiation can also interact with other global change factors such as ocean acidification and pollution. For example, ocean acidification can weaken the shells and exoskeletons of calcifying plankton like

Fig. 24.3   Image of a tropical reef. Australia. © Albert Calbet

coccolithophores and foraminifera, making them more susceptible to UV-induced damage. Similarly, pollutants such as microplastics and oil can alter the optical properties of seawater, potentially increasing the penetration of UV radiation and its harmful effects on marine life.

The broader ecological consequences of increased UV radiation on plankton are significant. Plankton form the base of the marine food web, and any disruption to their populations can ripple through the ecosystem. Changes in phytoplankton productivity and composition can affect the types and amounts of food available to zooplankton, which in turn impacts fish species that rely on zooplankton as a primary food source. This can have direct implications for commercial fisheries, which depend on the health and stability of these food webs for their livelihoods.

Changes in plankton communities due to increased UV radiation can also alter nutrient cycling and carbon sequestration. For example, a decline in phytoplankton productivity can reduce the efficiency of the biological carbon pump, where carbon is transported from the surface to the deep ocean. This can decrease the ocean's capacity to absorb carbon dioxide, contributing to further increases in atmospheric CO_2 levels and global warming.

UV radiation plays a crucial role in the sulfur cycle, particularly through its interaction with phytoplankton that produce dimethylsulfoniopropionate (DMSP). DMSP is a precursor to dimethyl sulfide (DMS), a compound that is released into the atmosphere and can contribute to cloud formation, which in turn influences the Earth's radiation budget. Increased UV-B radiation can reduce the production of DMSP by phytoplankton, leading to lower DMS emissions. This reduction in DMS can decrease cloud formation, potentially increasing the amount of solar radiation reaching the Earth's surface and contributing to further warming—a feedback mechanism that highlights the complex interplay between UV radiation, plankton, and climate.

25

The Silent Killers: Classic and Emerging Pollutants and Marine Plankton

When we read news about a catastrophic oil spill on our shores, we often think about pollution and its immediate consequences (though we tend to forget about plankton). However, these catastrophic events are merely the tip of the iceberg when it comes to ocean pollution. The health of marine ecosystems is intricately tied to the presence and concentration of pollutants in the ocean, many of which go unnoticed by the public. These pollutants, ranging from microplastics and heavy metals to agricultural runoff and industrial waste, accumulate over time and can have far-reaching effects on marine life and human health. Plankton, the foundation of marine food webs, are particularly vulnerable to these pollutants, which can disrupt their reproductive cycles and diminish their populations, ultimately affecting the entire marine ecosystem.

Pollutants can be broadly categorized into classic pollutants, which have been recognized and studied for decades, and emerging pollutants, which are more recently identified and are gaining attention due to their potential ecological impacts. Next, I will show some examples of the major effects of classic and emerging pollutants on plankton. I stress, however, that all these effects are dose-dependent and that open waters are usually below lethal and sublethal levels, unless a punctual catastrophe occurs. Yet, some particular coasts, bays, or estuaries may have pollutant concentrations that significantly affect plankton. What is more worrisome is that we simply do not know the effects of many of them on plankton.

Classic Pollutants: A Legacy of Contamination

Classic pollutants (Fig. 25.1) include substances such as heavy metals (e.g., mercury, lead, cadmium), hydrocarbons (e.g., petroleum and its derivatives), polychlorinated biphenyls (PCBs), and pesticides (e.g., DDT). These pollutants have been extensively studied, and their adverse effects on marine life, including plankton, are well-documented.

Heavy metals are persistent in the environment and can bioaccumulate in marine organisms, including plankton. For instance, mercury, once converted to methylmercury by microbial processes, becomes highly toxic. Methylmercury can bioaccumulate in phytoplankton, which are then consumed by zooplankton, and subsequently transferred up the food web. Studies have shown that exposure to mercury can impair the growth and reproduction of phytoplankton species such as *Phaeodactylum*

Fig. 25.1 Image of the shoreline of Jurong Island in Singapore, showing the multiple industries there. © Albert Calbet

tricornutum (a type of diatom). This impairment can reduce primary production, impacting the entire marine food web.

I recall, during my Ph.D., discovering that the mere presence of a copper ring in a filtration system completely halted the egg production of the copepod *Paracartia grani*. This finding highlighted the extreme sensitivity of these small but vital organisms to even trace amounts of pollutants.

Hydrocarbons (Fig. 25.2), especially those from oil spills, pose significant threats to plankton. The Deepwater Horizon oil spill in 2010, which released approximately 507 million liters (134 million gallons) of oil into the Gulf of Mexico, provided a stark example of the detrimental effects of hydrocarbons on marine ecosystems. Research indicated that polycyclic aromatic hydrocarbons (PAHs), a component of crude oil, can cause toxic effects on planktonic organisms. Studies on copepods revealed that exposure to PAHs resulted in reduced feeding rates, impaired development, and increased mortality. The disruption of copepod populations can have cascading effects on fish larvae and other marine organisms that depend on them for food.

The effects of hydrocarbons on plankton were also illustrated by the repercussion of the Exxon Valdez oil spill in 1989. Studies conducted in Prince William Sound revealed that oil contamination significantly reduced the abundance and diversity of phytoplankton and zooplankton.

Fig. 25.2 View from Singapore beach showing the tremendous amount of cargo ships waiting to enter the harbor. Singapore is one of the most active countries refining oil in the World. © Albert Calbet

Long-term monitoring showed that populations of key species, such as the copepod *Calanus marshallae*, were slow to recover, with ongoing repercussions for higher trophic levels.

PCBs and pesticides, such as DDT, are known for their persistence and bioaccumulative nefarious properties. Although the use of DDT has been banned in many countries, its residues still persist in marine environments. PCBs, used in various industrial applications, are still prevalent despite being banned or restricted in many areas. Both pollutants have been shown to cause reproductive and developmental issues in plankton. For instance, exposure to PCBs has been linked to deformities in copepod larvae, while DDT residues have been found to disrupt the reproductive cycles of various phytoplankton species. These disruptions can significantly alter plankton populations and their ecological roles.

Emerging Pollutants: New Challenges for Marine Life

Emerging pollutants encompass a wide range of substances, including pharmaceuticals, personal care products (PPCPs), microplastics, plasticizers, flame retardants, nanomaterials, and endocrine-disrupting chemicals (EDCs). These pollutants are often not fully regulated and can enter marine environments through various pathways, such as wastewater discharge, agricultural runoff, and atmospheric deposition.

Pharmaceuticals and PPCPs are increasingly detected in marine environments, raising concerns about their ecological impacts. Antibiotics, for example, can alter microbial communities by inhibiting the growth of sensitive species and promoting resistant strains. Research has shown that antibiotic residues can affect the structure and function of marine plankton communities. For instance, exposure to the antibiotic ciprofloxacin has been found to reduce the growth of marine cyanobacteria, which are crucial for nitrogen fixation in the ocean. Such disruptions can affect nutrient cycling and primary production.

Microplastics, tiny plastic particles less than 5 mm in diameter, have become ubiquitous in marine environments. They can originate from larger plastic debris that breaks down, as well as from products like cosmetics and synthetic clothing. Microplastics can be ingested by plankton, leading to physical and chemical impacts. Studies have demonstrated that, when offered at very high (unrealistic) concentration, copepods ingest microplastics, which can reduce their feeding efficiency and energy reserves, impairing their growth and reproduction. Additionally, microplastics can serve as vectors for other pollutants, such as heavy metals and hydrophobic organic compounds, which adhere to their surfaces and can be ingested along with the particles, exacerbating their toxic effects. In my humble opinion, microplastics do not pose an immediate threat to plankton at the present concentrations. However, further increases of these pollutants and associated chemicals may end up being detrimental for planktonic ecosystems.

Plasticizers, which give plastic goods their malleability, and *flame retardants*, used to reduce flammability, are two categories of emerging pollutants that enter the marine environment as plastic products degrade. Phthalates, like di(2-ethylhexyl) phthalate (DEHP), are known to disrupt the endocrine systems of marine organisms. Laboratory studies on planktonic copepods have shown that exposure to phthalates can lead to reproductive and developmental abnormalities. These substances interfere with normal hormonal functions, resulting in reduced fertility and altered growth patterns. Such disruptions can have cascading effects throughout the marine ecosystem, as plankton serve as a primary food source for many larger marine species. Moreover, flame retardants like polybrominated diphenyl ethers (PBDEs) have also been found to accumulate in marine plankton, further impacting their hormonal balance and overall health. The presence of these pollutants in plankton indicates a significant risk of bioaccumulation and biomagnification, potentially affecting higher trophic levels, including commercially important fish species.

The advent of nanotechnology has introduced *nanoparticles* into the environment, often stemming from industrial processes and consumer goods such as sunscreens, cosmetics, and toothpaste. These nanoparticles, including silver, titanium dioxide, and zinc oxide, are known for their antimicrobial properties but can be toxic to marine organisms. For instance, silver nanoparticles have been shown to cause oxidative stress and membrane damage in phytoplankton species like *Thalassiosira pseudonana*. Titanium dioxide nanoparticles, commonly found in sunscreens, have been shown to decrease the growth and chlorophyll content of the marine alga *Dunaliella tertiolecta*. Nanoparticles can also impair the photosynthetic efficiency of phytoplankton, leading to reduced primary production and potential disruptions in marine food webs.

Endocrine-disrupting chemicals (EDCs) are chemicals that can interfere with hormonal systems, even at low concentrations. These chemicals can enter marine environments through agricultural runoff, industrial discharge, and sewage effluents. EDCs such as bisphenol A (BPA) and certain pesticides can affect the development and reproduction of plankton. For example, studies have shown that exposure to BPA can disrupt the normal growth and reproductive cycles of copepods, leading to population declines. The presence of EDCs can also affect the behavior and physiology of plankton, potentially altering predator–prey interactions and overall ecosystem dynamics.

Bioaccumulation and Biomagnification

The concepts of bioaccumulation and biomagnification are crucial for understanding the broader impacts of pollutants on marine ecosystems. Bioaccumulation refers to the buildup of pollutants in an organism over time, often to toxic levels. Plankton, particularly phytoplankton, can absorb pollutants from their surrounding environment. As these pollutants accumulate in their tissues, they can reach concentrations much higher than those in the surrounding water. This process is particularly concerning for pollutants like PCBs and DDT, which are lipophilic (fat-loving) and can persist in the fatty tissues of organisms. When zooplankton feed on phytoplankton, they ingest these accumulated pollutants, which can then biomagnify as they move up the food web.

Biomagnification is the increase in concentration of a pollutant as it moves up through trophic levels. For instance, small fish (Fig. 25.3) that eat contaminated zooplankton will accumulate higher levels of the pollutant than that of zooplankton, and larger fish that eat smaller fish will accumulate even more. This can have severe implications for top

Fig. 25.3 Small fish, like sardines, accumulate relatively low amounts of pollutants compared to their larger counterparts, such as tuna

predators, including marine mammals and humans, who consume seafood. For example, methylmercury, which bioaccumulates in phytoplankton, can reach dangerously high levels in top predators like tuna and swordfish, posing health risks to both marine life and humans who consume these fish.

26

The Unseen Threats: Effects of Sound and Light Pollution on Plankton

In recent decades, human activities have introduced significant levels of sound and light pollution into marine environments, profoundly affecting marine life, including plankton. These pollutants are often overlooked compared to chemical contaminants, yet they pose serious threats to marine ecosystems. Plankton are particularly sensitive to changes in their environment. Understanding the impacts of sound and light pollution on plankton is crucial for assessing the broader ecological consequences and implementing effective conservation strategies.

Sound Pollution: A Growing Threat

Sound pollution, or anthropogenic noise, in marine environments primarily originates from shipping, industrial activities such as oil drilling, and naval sonar operations. This persistent noise disrupts the natural soundscape of the ocean, which many marine organisms rely on for communication, navigation, and predator–prey interactions. While plankton do not use sound for communication in the same way that larger marine animals such as whales and dolphins do, they can still be affected by acoustic disruptions.

The recent sparse studies have suggested that noise pollution can interfere with the behavior of zooplankton species. For instance, zooplankton, especially immature stages, can be killed by shots from seismic airguns. Moreover, sound pollution can induce physiological stress responses in plankton by disturbing their mechanoreceptors. These receptors help plankton detect changes in water movement and pressure, crucial for their survival. Disruption of these sensory systems can lead to altered swimming behavior, increased predation risk, and reduced reproductive success.

In addition to behavioral changes, noise pollution can cause physiological stress in plankton. Studies have indicated that elevated noise levels can increase metabolic rates in some planktonic species, leading to higher energy expenditure. For example, a study on the larval stages of marine oysters found that exposure to loud noises can result in elevated stress hormone levels and reduced growth rates. These physiological impacts can impair development and reduce the overall fitness of plankton populations. On the other hand, no effect on feeding was observed on early life stages of mussels and copepods by the presence of vessel noise.

Sound pollution can also interfere with the communication and echolocation abilities of larger marine animals, such as whales (Fig. 26.1), which feed on plankton. The disruption of these predators' ability to

Fig. 26.1 Whales are important consumers of large zooplankton and seem very affected by noise pollution. The image shows the tale of a whale in Greenland's waters. © Albert Calbet

locate and capture plankton can lead to changes in plankton populations and the structure of marine food webs. For instance, baleen whales rely on low-frequency sounds to locate dense patches of krill, a type of zooplankton. Increased noise levels can mask these sounds, making it more difficult for whales to find food and potentially leading to declines in their populations.

Light Pollution: Illuminating the Night Sea

Light pollution (Fig. 26.2), the excessive or misdirected artificial light typically associated with urbanization and coastal development, has also become a significant concern for marine environments. Plankton, much like terrestrial organisms, have evolved to follow natural light cues provided by the sun and moon. These cues regulate their circadian rhythms, influencing activities such as feeding, migration, and reproduction. Artificial light can mask these natural cues, leading to disrupted behavioral patterns. In the case of larval stages of marine invertebrates, such as sea urchins and crabs, exposure to artificial light has been shown to interfere with their settlement and metamorphosis processes. Similarly, continuous light exposure affects the feeding behavior and growth rates of marine dinoflagellates.

Moreover, many zooplankton undertake diel vertical migration, moving toward the surface at night to feed and descending to deeper waters during the day to avoid predators. Artificial lighting can disorient this migration pattern, leaving zooplankton exposed to predators. Phytoplankton and some species of zooplankton exhibit phototactic behavior, meaning they move toward or away from light sources. Artificial light pollution can cause these organisms to aggregate unnaturally around illuminated areas, leading to increased competition for resources and potential changes in community structure. Additionally, concentrated populations of plankton near artificial lights can attract more predators, further disrupting the balance of the marine ecosystem.

Fig. 26.2 Two examples of light pollution. Above: Shoreline of Hong Kong Island. Bellow: Marina Bay Sands, Singapore. Notice the strong illumination in both of them. © Albert Calbet

Mitigation and Management

Addressing the impacts of sound and light pollution on plankton requires a comprehensive and respectful approach toward the ocean, involving regulatory measures, technological advancements, public education, and ongoing research. It is key to remember that we do not own the ocean, we are mere visitors. As visitors, therefore, we should respect some basic rules, like we do on land when visiting a city.

Mitigating the effects of sound pollution involves multiple strategies, including reducing noise at the source, modifying shipping routes, and implementing "quiet zones" in sensitive marine areas. Technological advancements, such as quieter ship propellers and engines, can significantly reduce noise levels. Additionally, temporal restrictions on noisy activities during critical periods for marine life, such as breeding or migration seasons, can help protect vulnerable species. Research and monitoring are essential for understanding the long-term impacts of sound pollution on plankton. Studies using acoustic sensors and underwater microphones can help track noise levels and their effects on marine organisms. By combining these data with ecological studies, scientists can develop predictive models to assess the potential impacts of new marine infrastructure projects or changes in shipping patterns.

Addressing light pollution requires coordinated efforts to reduce artificial light at night, especially in coastal areas. This can involve using shielded lighting to minimize light spill, adjusting the timing and intensity of lighting, and promoting the use of "dark sky" principles. For offshore platforms and ships, using red or amber lights, which are less disruptive to marine life, can help mitigate the effects of light pollution. Implementing policies that regulate light emissions and encourage the use of environmentally friendly lighting can significantly reduce the impact on marine ecosystems. Public awareness campaigns and community engagement are also crucial in promoting best practices for lighting in coastal areas.

27

Plankton and Global Fisheries

If Bob Marley had been more interested in marine biology than in personal relationships, his famous song "No Woman, No Cry" might have been "No Plankton, No Fish." Try to sign it, it is kind of catchy. Seriously speaking, the truth is that global fisheries are intrinsically linked to the health and abundance of plankton populations. The complex relationships between plankton and fish species form the backbone of marine ecosystems, influencing fish stocks, marine biodiversity, and the economic stability of coastal communities. However, overfishing and environmental changes pose significant threats to these delicate relationships, leading to shifts in plankton populations that can have cascading effects on fisheries and marine ecosystems.

Plankton and Fisheries: An Intricate Relationship

The health of global fisheries is deeply intertwined with the dynamics of plankton populations. Many commercially important fish species rely on plankton at various stages of their lifecycles. For instance, larval and juvenile stages of fish such as herring, mackerel (Fig. 27.1), sardines, and

A. Calbet, *Plankton in a Changing World*,
https://doi.org/10.1007/978-3-031-76121-8_27

Fig. 27.1 Mackerel painting. © Albert Calbet

anchovies depend on zooplankton as a primary food source. The availability and abundance of zooplankton directly influence the survival rates and growth of these young fish, which in turn affects adult fish populations and fishery yields.

Phytoplankton also play a crucial role in supporting fisheries. As primary producers, they form the base of the marine food web. Phytoplankton blooms can trigger zooplankton blooms, providing a rich food source for fish larvae. However, changes in phytoplankton composition and abundance, driven by environmental factors and overfishing, can disrupt these food webs.

The Case of *Calanus finmarchicus* and *Calanus helgolandicus*

A significant example of how changes in plankton populations affect fisheries is the replacement of the copepod species (Fig. 27.2) *Calanus finmarchicus* by *Calanus helgolandicus* in the North Atlantic. *C. finmarchicus* is a cold-water species that serves as a crucial food source for many fish species, including Atlantic cod. This copepod species is particularly abundant in the colder waters of the North Atlantic and is known for its high lipid content, which is essential for the energy needs of fish larvae.

However, with rising sea temperatures due to climate change, there has been a noticeable shift in the distribution of *C. finmarchicus*. Warmer

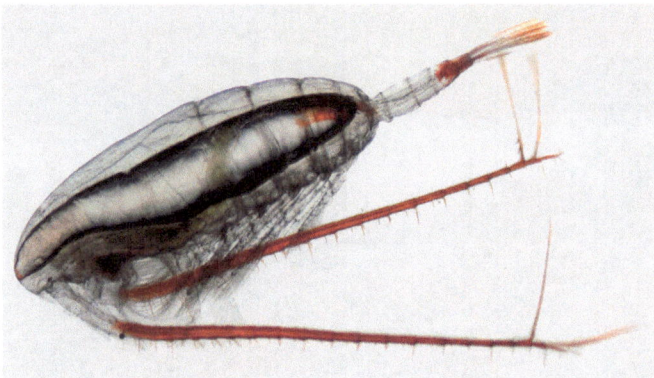

Fig. 27.2 The genus *Calanus* includes several species that, while similar in appearance, possess different nutritional compositions. In the picture a *Calanus hyperboreus*, a high-Arctic species. © Albert Calbet

waters have facilitated the northward expansion of *C. helgolandicus*, a species that prefers slightly warmer conditions. This shift has significant implications for Northern fisheries, particularly those targeting cod.

Atlantic cod larvae depend heavily on the lipid-rich *C. finmarchicus* during their early life stages. The replacement of this species by the less nutritious *C. helgolandicus* means that cod larvae have access to a lower quality food source, which can lead to reduced growth rates, lower survival rates, and ultimately, lower recruitment into the adult population. This shift can have severe consequences for the productivity and sustainability of Northern cod fisheries.

Overfishing and Trophic Cascades

Overfishing not only directly depletes fish populations (Fig. 27.3) but also disrupts the intricate food webs that support marine ecosystems. One of the most significant ways overfishing affects plankton is through trophic cascades. Trophic cascades occur when changes at one trophic level cause a series of effects across other levels. For example, the overfishing of large predatory fish, such as tuna and cod, can lead to an increase in the populations of their prey, which often include smaller fish and

Fig. 27.3   Fish market, Valdivia (Chile). © Albert Calbet

invertebrates that consume zooplankton. This increase in zooplanktivo-rous species can lead to a decline in zooplankton populations, which in turn can affect phytoplankton dynamics.

A well-documented example of a trophic cascade involves the overfish-ing of large predatory fish in the North Atlantic. Studies have shown that the depletion of cod populations led to an increase in smaller fish and invertebrates, such as jellyfish and mesopelagic fish, which prey heavily on zooplankton. The resulting decline in zooplankton abundance affected the entire marine ecosystem, altering nutrient cycling and primary pro-duction. In the Baltic Sea, the overfishing of cod has been linked to an increase in sprat and herring populations. These small pelagic fish con-sume large quantities of zooplankton, leading to a decrease in zooplank-ton biomass. This reduction in zooplankton has been associated with increased phytoplankton biomass and more frequent harmful algal blooms, which negatively impact water quality and marine life. In the Black Sea, overfishing of top predators such as bonito and mackerel led

Fig. 27.4   Jellyfish. *Chrysaora* sp. © Albert Calbet

to an increase in jellyfish populations. Jellyfish (Fig. 27.4) are voracious feeders on zooplankton, and their proliferation resulted in a significant decline in zooplankton populations. This shift in the food web structure contributed to the collapse of commercial fish stocks and the degradation of the marine ecosystem. Finally, a theoretical positive relationship has been suggested between whales and krill in Antarctic waters. A decrease in the number of whales led to reduced recycling of iron, a crucial element for phytoplankton growth, which in turn supports krill populations. These are a few of many examples of overfishing effects on plankton populations.

Required Actions

Addressing the impact of overfishing on plankton and marine ecosystems requires a multifaceted approach. Implementing sustainable fishing practices, such as setting catch limits based on scientific assessments, protecting critical habitats, and reducing bycatch, is essential for maintaining healthy fish populations and the marine food web.

Marine protected areas can play a significant role in conserving marine biodiversity and supporting the recovery of overfished populations. They provide refuges where fish populations can grow and reproduce without the pressures of fishing, leading to spillover benefits for surrounding areas.

Monitoring and research are also crucial for understanding the complex interactions between plankton and fisheries. Long-term studies, such as those conducted by the Continuous Plankton Recorder survey (Plymouth, UK) that started in 1931 or that of Narragansett Bay (USA) beginning in 1957, provide valuable data on plankton trends and help identify the impacts of environmental changes and human activities on marine ecosystems.

28

Predicting the Future of Our Seas and Ocean Ecosystems: Reality or Chimera?

Many of you may have come across alarmist news regarding the future of seas and ocean, particularly by the end of this century (2100). Typically, this information, extracted and often misinterpreted from scientific articles by journalists with limited expertise in the field, tends to be alarmist, painting a rather bleak picture for emblematic or commercially significant communities like corals, gorgonians, fish, whales, etc. However, few consider the fate of plankton, a crucial group of marine organisms increasingly recognized as essential for understanding the future of ocean life.

Have you ever wondered where this information originates? In this chapter, I will attempt to explain it. Let's start by considering the timeline for the expiration of marine ecosystems. The year 2100 is frequently used as a benchmark in climate change predictions, where all forecasts converge. Most of us, except the youngest, may not witness this date. So, why should we care? Because 2100 seems distant and does not immediately incite concern, using such a long-term prediction may be a mistake. It distances us from the imminent and progressive changes we are already witnessing. Nevertheless, I understand the political and scientific rationale behind it. What sense would have to predict the ocean fate by the end of next year?

Returning to the Main Topic, How Do We Predict Future Events?

Predictions rely on mathematical models that extrapolate current variables to a more or less distant future, using equations of varying complexity. Plankton models incorporate a range of factors such as nutrient availability, light intensity, and temperature, each of which plays a critical role in influencing phytoplankton growth. For instance, nutrient availability, particularly nitrogen and phosphorus, determines the rate at which phytoplankton populations expand or contract. Similarly, light intensity, which is affected by water clarity and seasonal variations, is crucial for photosynthesis. Temperature also significantly impacts phytoplankton, with some species thriving in warmer waters while others struggle. Accurately quantifying these factors involves significant uncertainties and variabilities, posing challenges for models. Moreover, for these models to be accurate, they also require reliable data and thorough validation on how organisms react to specific stimuli. This is where the most significant problem arises. To feed the models, we need data on organism abundance, production, growth, inter-species relationships, and the impact of various physical-chemical variables, among others. No current model has the capacity to handle numerous species simultaneously, necessitating grouping by functions, sizes, etc. This simplification poses a challenge as it leads to information loss. On the other hand, overly complex models may collapse or yield inaccurate conclusions.

Just as a glimpse of the various modeling approaches, each with strengths and limitations, here you have some of them:

Nutrient-Phytoplankton-Zooplankton (NPZ) Models: These relatively simple models (Fig. 28.1) represent the interactions between nutrients, phytoplankton, and zooplankton. By tracking these interactions, NPZ models estimate CO_2 removal by considering phytoplankton growth and organic matter sinking.

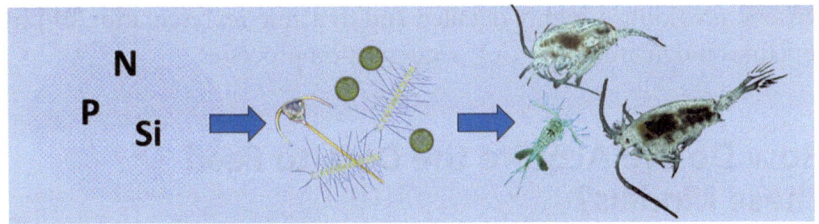

Fig. 28.1 NPZ models, despite their simplicity and early origine, are still used nowadays. Albert Calbet

Earth System Models (ESMs): ESMs offer a more comprehensive view by incorporating atmospheric, oceanic, and terrestrial processes. ESMs can simulate phytoplankton dynamics alongside global circulation patterns, providing a holistic view of CO_2 removal within the Earth's climate system.

Size-Structured Models: These models account for the size diversity within phytoplankton communities, crucial for understanding different size classes' contributions to CO_2 sequestration.

Plankton Functional Type (PFT) Models: PFT models categorize plankton into functional groups based on their ecological roles, allowing for a more nuanced understanding of their contributions to carbon cycling.

Biogeochemical Models: These models integrate physical, chemical, and biological processes to simulate biogeochemical cycles, providing detailed insights into nutrient cycling and carbon export.

Regional Ocean Modeling Systems (ROMS): ROMS simulate physical ocean processes and their interactions with biogeochemical cycles, useful for studying specific regions.

Individual-Based Models (IBMs): IBMs simulate individual organism behaviors and interactions, capturing fine-scale processes influencing CO_2 removal.

Coupled Physical-Biological Models: These models integrate physical oceanography with biological processes, essential for understanding how physical conditions influence phytoplankton dynamics.

Digital Twins: They offer a promising cutting-edge approach to simulating marine ecosystems. These virtual models can integrate real-time data from ship-based surveys, satellite observations, and autonomous sensors, providing a highly detailed and dynamic representation of phytoplankton dynamics and CO_2 sequestration processes.

How Do We Acquire the Data to Feed These Models?

Focusing simply on the biological aspect, the diversity of functions in plankton alone is overwhelming. Various approaches exist to obtain data, such as laboratory experiments investigating the impact of temperature on a representative species. However, strain differences, long-term adaptations, synergies, and interactions with other factors complicate the interpretation of these experiments. Another approach involves isolating entire natural communities and studying the effects of changing parameters, like temperature, through microcosm or mesocosm experiments. Yet, these experiments face challenges in controlling multiple interactions and may yield results dependent on the specific community and time period studied.

Continuous monitoring of specific areas, particularly during seasons with conditions approaching those expected in the future, is perhaps the best way to understand and predict the fate of marine communities. However, this approach yields localized results, and there's a risk of overlooking long-term adaptations, migrations, or invasions of foreign species. Additionally, such studies demand significant human effort and face funding constraints, with low scientific productivity.

Despite These Challenges, Should We Give Up?

Science has always progressed gradually, with major discoveries interspersed among routine contributions. However, we are at the beginning of a new era, and the demand for solutions exceeds our ability to respond. Moreover, the study of the ocean has not been a priority for most governments, and funding remains limited. With the current pace of climate and non-climate changes (such as pollution and other anthropogenic impacts), marine communities might be evolving faster than our ability to understand and predict them reliably.

We, as a species, find ourselves in a unique situation with the ability to improve the planet and future generations' well-being, yet it seems there is a lack of desire or firm intention to do so. Uncertainty about the future is greater than ever, and despite having the means to (at least partially) illuminate our path, we often choose to run forward in the dark, hoping for luck.

Regarding the future of the ocean and their natural communities, despite negative perspectives, I remain optimistic. Profound changes will occur, but life will find a way to adapt. New communities may not align with human preferences and needs, accustomed to viewing the sea as a source of resources, food, transport, and nowadays even water. The changes have commenced, and even if we halt emissions, pollution, or overexploitation of fisheries, the ocean will still undergo profound alterations. However, preventing drastic ecosystem transformations or slowing their pace remains within our control.

29

The Plankton Time Machine: A Journey Through Geological Eras

Imagine stepping into a time machine that opens a window into the history of our planet. Remarkably, such a time machine exists in the form of tiny plankton. These minuscule entities have been silent witnesses to Earth's dramatic transformations over millions of years. By studying fossilized plankton, scientists have assembled invaluable insights into prehistoric climates, oceanic conditions, and even the atmospheric composition of ancient Earth.

Plankton: The Ancient Drifters

Plankton have not only played an essential role in marine ecosystems but have also significantly influenced geological and climatic eras. Phytoplankton, for example, are prolific photosynthesizers and have been instrumental in regulating atmospheric oxygen and carbon dioxide levels. This has implications not only for today's climate but also for the primordial atmosphere, which was vastly different from what we experience now.

The history of plankton dates back to the Proterozoic Eon, around 2 billion years ago, when single-celled algae began to photosynthesize, releasing oxygen as a by-product. This early activity contributed to the Great Oxidation Event, drastically altering the planet's atmospheric composition. Fast forward to the Cambrian Period, approximately 540 million years ago, when plankton diversified dramatically. Fossils from this era show a proliferation of diatoms and radiolarians, and even the origin of copepods, revealing a rich tapestry of early marine biodiversity.

Fossilized Plankton: Clues from the Past

Fossilized plankton serve as critical proxies for understanding Earth's historical climate. One of the most significant groups of fossilized plankton is the foraminifera (Figs. 29.1 and 29.2). These tiny, shell-bearing organisms have left behind extensive fossil records in marine sediments. By

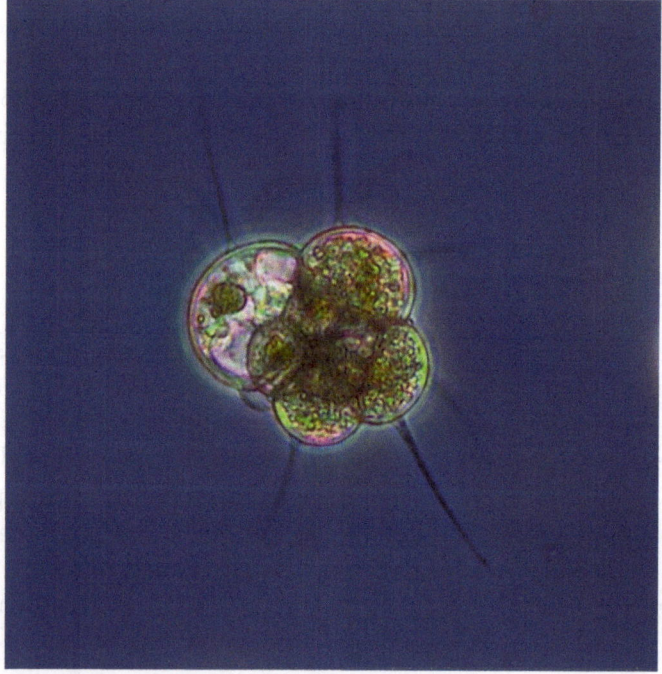

Fig. 29.1 Planktonic foraminifera. © Albert Calbet

Fig. 29.2 Nummulites were large lenticular benthic foraminifera common in the fossil record. Nummulites were very large, up to 6 cm in diameter and were common in Eocene to Miocene marine rocks. © Albert Calbet

analyzing the isotopic composition of foraminifera shells, scientists can infer temperature changes and oceanic conditions over millennia. For example, the ratio of oxygen isotopes (O-16 and O-18) in these shells varies with water temperature, allowing researchers to reconstruct past climates with remarkable accuracy.

Another group, coccolithophores, produces calcium carbonate plates called coccoliths that accumulate on the ocean floor. During periods of high biological activity, vast deposits of coccoliths formed chalk layers, such as those found in the famous White Cliffs of Dover. By examining these formations, scientists can determine periods of high carbon sequestration, offering insights into how ancient marine ecosystems responded to shifts in atmospheric CO_2 levels.

Diatoms (Fig. 29.3), another type of phytoplankton, produce silica-based cell walls that also accumulate as sediment. These sediments, known as diatomaceous earth, provide clues about nutrient cycles and ocean productivity over geological timescales. Diatom fossils are particularly useful for studying past ice ages because their abundance and distribution are influenced by sea ice cover and nutrient availability.

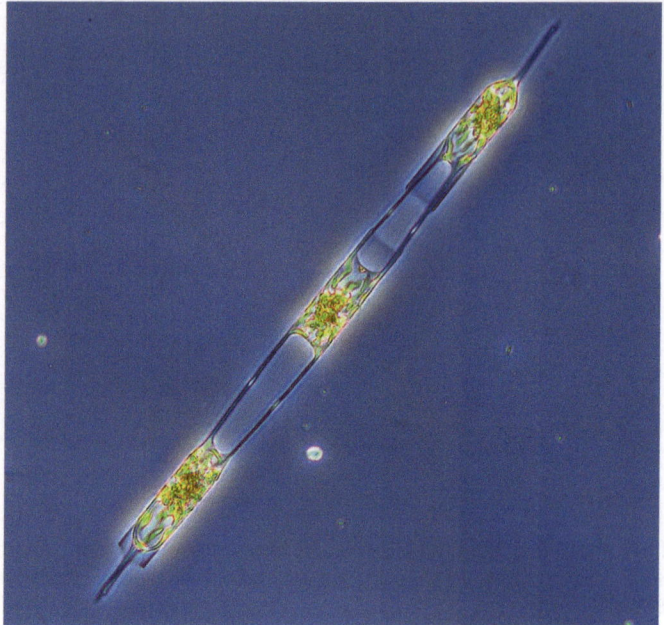

Fig. 29.3. Marine diatom (*Hemiaulus* sp.). © Albert Calbet

Decoding Earth's Climatic History

The study of fossilized plankton has illuminated key periods in Earth's climatic history. For instance, during the Cretaceous Period, approximately 145 to 66 million years ago, the Earth experienced a "greenhouse" climate with higher global temperatures and elevated sea levels. Plankton fossils from this time reveal increased primary productivity, suggesting a thriving marine ecosystem. However, this period also saw massive volcanic eruptions that released large amounts of CO_2, contributing to the intense greenhouse conditions.

Conversely, the Pleistocene Epoch, spanning from about 2.6 million to 11,700 years ago, was characterized by repeated glacial cycles. Plankton fossils from this epoch indicate lower ocean temperatures and significant ice cover. By studying the changes in plankton communities, scientists can piece together the timing and extent of glacial advances and retreats, providing a clearer picture of Earth's climatic oscillations.

Moreover, plankton fossils have shed light on mass extinction events. The end-Permian extinction, around 252 million years ago, saw the disappearance of approximately 90% of marine species. Plankton records show a dramatic decline in biodiversity, followed by a slow recovery during the Triassic Period. Similarly, the end-Cretaceous extinction, famously linked to the demise of the dinosaurs, also had profound effects on marine plankton communities. The rapid deposition of iridium-rich layers, coupled with abrupt shifts in plankton populations, supports the asteroid impact hypothesis for this mass extinction.

Implications for Contemporary Climate Science

Understanding the history of plankton and their fossil records is not merely an academic exercise; it has profound implications for contemporary climate science. As we face unprecedented levels of CO_2 emissions and global warming, plankton offer vital lessons on how Earth's systems respond to such changes. Past episodes of elevated CO_2 and their associated effects on marine life provide analogs for predicting future scenarios.

By studying these ancient drifters, we gain not only a window into Earth's past but also a guide for navigating the environmental challenges of the present and future. The silent witnesses of our planet's history, plankton, continue to teach us about the intricate and enduring relationships between life and climate.

Part IV

Case Studies and Regional Perspectives

Part IV

Case Studies and Personal Perspectives

30

The Effects of Global Change on Plankton in Polar Regions

Despite my aversion to cold (I prefer my warmer Mediterranean), I have always been fascinated by the polar regions. I have had the opportunity to visit both the Arctic and Antarctica for work on several occasions, and the views are absolutely breathtaking. Each time that I gazed upon those pristine white landscapes, I felt incredibly fortunate—not only for being there but also because I knew that in a few decades, those views might be drastically different. Indeed, the polar regions are among the most vulnerable ecosystems to global environmental changes. While both the Arctic and Antarctic are experiencing significant environmental changes, there are key differences in their ecosystems and responses to global change. The Arctic is an ocean surrounded by continents, which means that changes in terrestrial input, such as freshwater runoff from melting glaciers and riverine discharge, can significantly influence its marine ecosystems. In contrast, the Antarctic is a continent surrounded by the Southern Ocean. Both regions are experiencing rapid and profound shifts due to rising temperatures, changing ice cover, and shifting ocean currents. Plankton are particularly sensitive to these changes. The impacts on plankton in these regions have cascading effects on the entire ecosystem, affecting everything from primary production to the survival of top predators like whales, seals, and seabirds.

A. Calbet, *Plankton in a Changing World*,
https://doi.org/10.1007/978-3-031-76121-8_30

Temperature Rise and Its Impact on Polar Plankton

Global warming has led to significant increases in sea surface temperatures, particularly in polar regions. The Arctic is warming at twice the rate of the global average, a phenomenon known as Arctic amplification. This rapid warming has profound effects on plankton communities.

Phytoplankton, which (as all organisms) require specific temperature ranges to thrive, are directly affected by rising temperatures. Warmer waters can lead to shifts in phytoplankton species composition, favoring those adapted to higher temperatures and causing the decline or extinction of species highly specialized for near-freezing waters. For instance, diatoms, which are dominant in cold waters, may be replaced by smaller phytoplankton such as flagellates and picoplankton in warmer conditions. A particular genus of concern is *Phaeocystis* (Fig. 30.1), which is inedible by most zooplankters. This shift can reduce the overall productivity of the ecosystem because diatoms are highly efficient at carbon fixation and form the base of the food web for many larger organisms.

In the Antarctic, the Southern Ocean's temperature rise is altering phytoplankton blooms. The timing and magnitude of these blooms are shifting, which affects the entire food web. For example, earlier and more intense phytoplankton blooms are occurring, driven by melting sea ice and increased light penetration into the water column. This can benefit some species but disrupt the lifecycles of others that are adapted to specific bloom periods.

Sea Ice Decline and Plankton Dynamics

The decline of sea ice (Fig. 30.2) due to global warming is one of the most visible changes in the polar regions. Sea ice plays a crucial role in polar ecosystems, particularly for plankton. It provides a unique habitat for ice-associated algae, which are a critical food source for zooplankton during the winter months when other sources are scarce.

In the Arctic, the reduction in sea ice cover leads to longer ice-free periods, which in turn alters the timing and distribution of

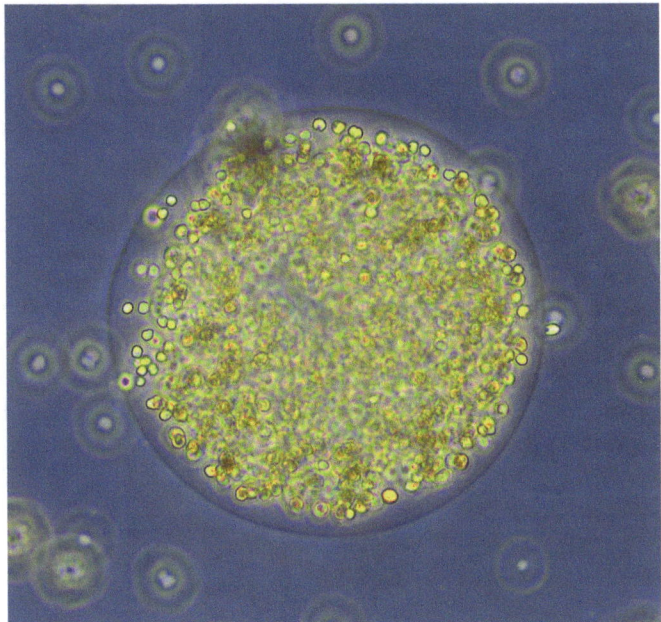

Fig. 30.1 *Phaeocystis* sp. colony. © Albert Calbet

Fig. 30.2   Sea ice. Arctic waters. © Albert Calbet

phytoplankton blooms. As sea ice melts earlier in the spring, the increased light availability promotes earlier phytoplankton blooms. However, these blooms may not always align with the lifecycles of zooplankton, such as copepods, which rely on the timing of these blooms for their reproduction and growth. This mismatch can lead to reduced food availability for zooplankton and subsequently impact the species that feed on them, including fish, seabirds, and marine mammals.

In the Antarctic, the shrinking and thinning of sea ice similarly disrupts the habitat for ice algae. The reduction in sea ice has led to changes in the species composition and abundance of phytoplankton. For example, the abundance of large diatoms, which thrive under the ice, has decreased, while smaller phytoplankton species have increased. These algae are a crucial food source for krill during winter months when open water phytoplankton is scarce. Krill scrape the algae off the ice, allowing them to sustain themselves in the harsh winter conditions.

Changes in Ocean Circulation and Nutrient Availability

Climate change is also altering ocean circulation patterns, which affects the distribution of nutrients critical for plankton growth. In the Arctic, the retreat of sea ice and warming temperatures are altering the stratification of the water column, impacting nutrient mixing and availability.

One significant change is the increased inflow of Atlantic water into the Arctic Ocean, bringing warmer, nutrient-rich waters. While this can enhance phytoplankton productivity in some areas, it can also lead to stratification that limits nutrient upwelling in others. This uneven nutrient distribution can cause shifts in plankton communities, with potential impacts on the entire food web. Moreover, the seasonal melting of the Arctic's sea ice is crucial to maintain the global circulation patterns of the ocean and to fuel the Gulf Stream. Recent predictions suggest that in few years/decades this current could reach a full stop, which would have devastating consequences for the whole planetary climate. Without the Gulf Stream, Europe and North America would experience significantly colder winters, disrupting agriculture and leading to food shortages. Marine ecosystems would also be thrown into chaos, affecting fish populations that many communities rely on for sustenance. Moreover, a halt in the Gulf Stream could accelerate sea-level rise along the eastern coast of the United States and contribute to more extreme weather patterns globally, posing profound challenges to both human societies and natural environments.

In the Southern Ocean, changes in the Antarctic Circumpolar Current due to warming temperatures and shifting wind patterns are affecting nutrient distribution. This can influence the productivity and composition of phytoplankton communities. Changes in iron availability, a critical nutrient in the Southern Ocean, can significantly impact phytoplankton blooms and their composition.

Plankton Changes in Polar Regions

Several scientific studies (Fig. 30.3) have documented the impacts of global change on polar plankton. In the Arctic, besides the deleterious effects of pollution from the Northern hemisphere, research highlighted the consequences of sea ice decline on the timing of phytoplankton

Fig. 30.3 The Spanish research vessel BIO Hesperides in the Arctic Ocean. © Albert Calbet

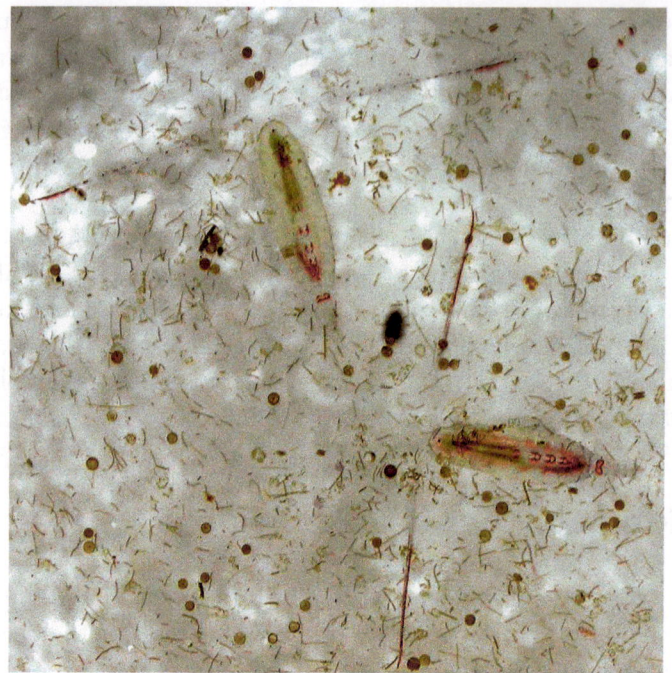

Fig. 30.4 *Calanus glacialis* in a diatom bloom. Arctic waters. © T. G. Nielsen

blooms and zooplankton mismatches (Fig. 30.4). It was found that earlier ice melt led to earlier phytoplankton blooms, which did not coincide with the peak reproductive period of zooplankton, resulting in reduced food availability for higher trophic levels.

In the Antarctic, studies on long-term changes in phytoplankton communities in response to sea ice variability observed a decline in large diatoms (and krill) and an increase in smaller phytoplankton species, correlating these changes with reduced sea ice cover and increased light availability, among other factors. Furthermore, research on ocean acidification demonstrated the vulnerability of pteropods in Polar systems. Experiments showed that shell dissolution occurred rapidly under acidified conditions, highlighting the potential for significant impacts on these organisms and the food web.

31

The Effects of Global Change on Plankton in Tropical Regions

Tropical regions, characterized by warm waters and high biodiversity, are experiencing significant shifts due to global environmental changes. Rising sea temperatures, ocean acidification, and altered precipitation patterns profoundly impact these marine ecosystems. The effects on plankton populations in tropical regions have cascading impacts on the entire ecosystem, influencing everything from primary production to the survival of fish, marine mammals, and seabirds. This chapter explores the effects of global change on tropical plankton, supported by scientific studies and examples.

Rising Sea Temperatures and Plankton Dynamics

Global warming has led to substantial increases in sea surface temperatures. This is particularly important in tropical regions, because many species inhabiting these ecosystems are already at their tolerance limit.

© The Author(s), under exclusive license to Springer Nature Switzerland AG 2024
A. Calbet, *Plankton in a Changing World*,
https://doi.org/10.1007/978-3-031-76121-8_31

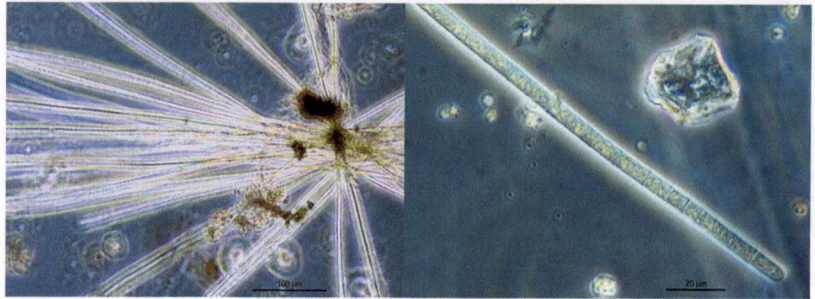

Fig. 31.1 *Trichodesmium*. Left: colony. Right: detail of one fiber. © Sergi Rodriguez López

Therefore, warmer waters significantly affect plankton communities, leading to shifts in species composition, abundance, and productivity.

For example, a study in the Western Pacific Ocean observed a significant increase in the abundance of the nitrogen-fixing cyanobacterium *Trichodesmium* (Fig. 31.1) in response to rising sea temperatures. This shift not only alters the nutrient dynamics in the region but also affects the entire food web, as *Trichodesmium* supports different zooplankton species compared to diatoms.

Ocean Acidification and Its Effects on Plankton

In tropical waters, calcifying phytoplankton, such as coccolithophores and foraminifera, are particularly vulnerable. These organisms play a crucial role in the carbon cycle, and their decline can disrupt marine ecosystems. Studies have shown that acidified conditions reduce the growth and calcification rates of coccolithophores, leading to shifts in phytoplankton community structure. For instance, research in the Great Barrier Reef in Australia, and the Coral Triangle, a region in the western Pacific Ocean, has shown that acidification negatively impacts the growth and shell formation of foraminifera, crucial components of the reef ecosystem.

Changes in Precipitation and Freshwater Input

Global climate change is also altering precipitation patterns and freshwater input into tropical marine environments. Increased rainfall and river discharge can lead to changes in salinity and nutrient availability, significantly impacting plankton communities.

For example, the Amazon River, one of the largest sources of freshwater input to the Atlantic Ocean, influences the nutrient dynamics and plankton productivity in the region. Increased rainfall and runoff can enhance the supply of nutrients such as nitrogen and phosphorus, leading to phytoplankton blooms. However, excessive nutrient input can also lead to eutrophication, causing harmful algal blooms that disrupt marine ecosystems, from copepods (Fig. 31.2) up, and threaten fish populations. Research in the Amazon River plume has documented shifts in phytoplankton communities in response to varying freshwater input. During periods of high discharge, the abundance of diatoms increases, supporting a different zooplankton community compared to periods of low discharge, when cyanobacteria dominate. These shifts have significant implications for the broader marine food web. In the South China Sea, rising sea temperatures and changing monsoon patterns are altering the timing and magnitude of phytoplankton blooms. These changes are

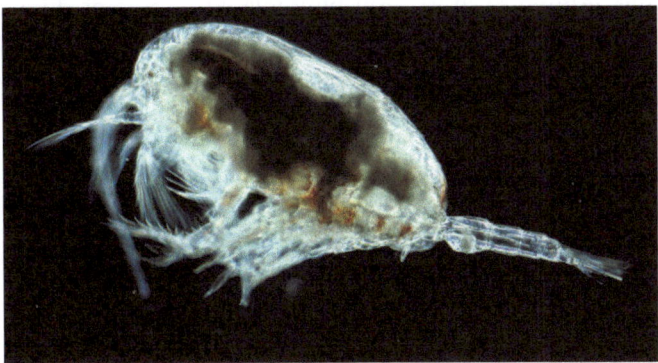

Fig. 31.2 The genus *Clausocalanus* is a typical representative of tropical and temperate ecosystems. © Albert Calbet

affecting the entire marine food web, from zooplankton to fish and marine mammals.

Tropical Cyclones and Their Impact on Plankton

Tropical cyclones, which are becoming more intense due to global warming, have immediate and long-term effects on plankton communities. These storms can cause physical mixing of the water column, bringing nutrients from the deep ocean to the surface and promoting phytoplankton blooms. However, the associated heavy rainfall and runoff can also introduce large amounts of sediments and pollutants, negatively impacting water quality and plankton health.

For instance, studies in the Gulf of Mexico have shown that hurricanes can lead to short-term increases in phytoplankton productivity due to nutrient upwelling. Yet, the subsequent runoff and sedimentation can lead to declines in water quality, reducing light penetration and negatively affecting phytoplankton growth. The balance between these positive and negative effects varies depending on the intensity and frequency of the storms. Another example comes from the Indian Ocean, where increased frequency and intensity of cyclones are impacting phytoplankton productivity. Studies have documented how cyclones enhance nutrient mixing, leading to short-term phytoplankton blooms. However, the long-term effects include changes in species composition and declines in overall productivity due to increased sedimentation and reduced light availability.

32

Effects of Global Change on Plankton in Temperate Regions

Temperate regions, characterized by moderate climates and distinct seasonal variations, host diverse and dynamic plankton communities that play a crucial role in marine ecosystems. Plankton in these regions are highly responsive to changes in environmental conditions, including temperature fluctuations, nutrient availability, and light cycles, making them particularly vulnerable to the impacts of global change. This chapter explores the effects of global change on temperate plankton, focusing on how different seasons influence these impacts. The discussion is supported by scientific studies and examples of specific species.

Seasonal Variability and Plankton Dynamics

In temperate regions, plankton dynamics are closely tied to seasonal cycles (Fig. 32.1). Spring and fall are typically the periods of highest phytoplankton productivity, driven by increased light availability and nutrient upwelling. During winter, lower temperatures and reduced light limit phytoplankton growth, while summer often sees a decline in productivity due to nutrient depletion and stratified water columns.

© The Author(s), under exclusive license to Springer Nature Switzerland AG 2024
A. Calbet, *Plankton in a Changing World*,
https://doi.org/10.1007/978-3-031-76121-8_32

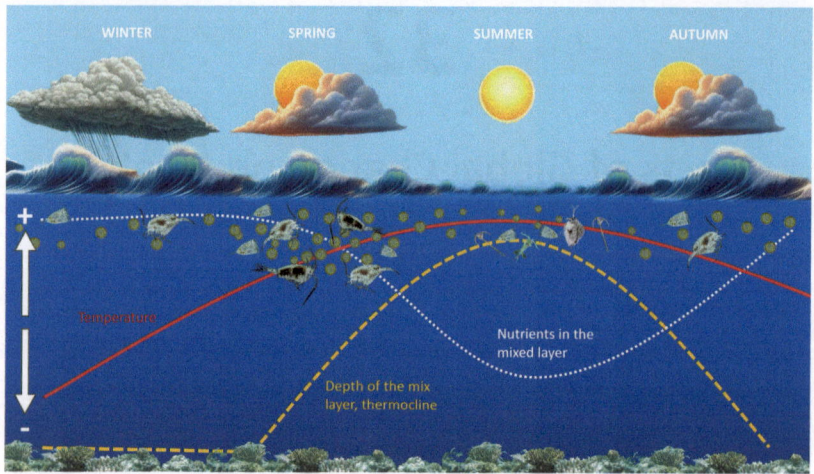

Fig. 32.1 In a temperate system, plankton exhibit limited growth in nutrient-rich but light-poor winter waters; spring blooms fueled by increasing light and temperature; summer stratification leading to nutrient depletion in the surface layer and a shift to small algae; and autumn storms breaking the thermocline, potentially triggering smaller secondary blooms before winter returns. © Albert Calbet

Global change is altering these seasonal patterns, with significant implications for plankton communities. Rising sea temperatures, changes in precipitation patterns, and altered ocean circulation are all contributing to shifts in plankton dynamics.

Rising Sea Temperatures and Seasonal Shifts

Rising sea temperatures are one of the most direct effects of global change on plankton. In temperate regions, warmer temperatures can lead to shifts in the timing and intensity of seasonal plankton blooms. For instance, earlier onset of spring warming can trigger earlier phytoplankton blooms. While this might seem beneficial, it can lead to mismatches in the timing of plankton availability and the lifecycles of zooplankton and higher trophic levels.

A notable example is the shift in the timing of the spring phytoplankton bloom in the North Atlantic and that of *Calanus finmarchicus*. This shift has disrupted the feeding patterns of *C. finmarchicus*, which rely on the spring bloom for food during their critical growth periods. The resulting mismatch can reduce zooplankton survival and impact the entire food web, including fish populations that depend on zooplankton for food. In the Mediterranean Sea, sea surface temperatures have risen by approximately 1.5 °C over the past century. This warming has led to earlier and longer-lasting phytoplankton blooms, which can result in shifts in the entire marine food web. This increase of temperature is also associated with smaller species of lower nutritional quality, which seems to be affecting local fisheries of sardine and anchovy. In the English Channel, long-term monitoring has revealed shifts in phytoplankton species composition due to rising sea temperatures. Species such as *Pseudo-nitzschia delicatissima*, a diatom associated with harmful algal blooms, have become more prevalent, posing risks to marine life and human health.

Heatwaves and Their Consequences

Marine heatwaves—prolonged periods of abnormally high sea temperatures—are becoming more frequent and intense due to global warming. These heatwaves can have devastating effects on plankton communities, leading to sudden and drastic changes in species composition and productivity.

Toward the end of the summer in 1999, the Mediterranean was struck by a marine heatwave that reached depths of up to 50 meters, causing significant mortality among gorgonians and other benthic organisms. Similar levels of mortality were again recorded in August 2022 when a marine heatwave caused temperatures to rise by +4 to +6 °C above average in the Mediterranean. In the equatorial Pacific, the extreme warm conditions of the 2016 El Niño led to decline of about 40% in surface chlorophyll, which was associated with a nearly total collapse in diatoms. In the Eastern North Atlantic, strong to severe marine heatwaves were detected as of late June 2023. These heatwaves resulted in a significant decrease in chlorophyll-a concentration, with reductions of up to 50–60%.

Zooplankton are no-immune to heatwaves. For instance, in the 2003 heatwave in Europe resulted in a significant decline in copepod (Fig. 32.2)

Fig. 32.2 There are several research groups, including ours, currently conducting research on heatwaves and long-term effects of temperature on *Acartia* and *Paracartia* species. In the picture a female of *Paracartia grani*. © Albert Calbet

abundance in different areas, particularly in the Mediterranean. Experimental evidence demonstrated that besides the direct effect of temperature, zooplankton consuming "heatwave" phytoplankton attained lower community biomass than those fed with phytoplankton from "constant warming" or "ambient" conditions. Despite similar total heat input, phytoplankton exposed to "heatwave" conditions contained lower concentrations of carbon (C), nitrogen (N), phosphorus (P), and fatty acids compared to those grown in "constant warming" conditions. This highlights the critical impact of marine heatwaves on the nutritional quality of phytoplankton, which directly affects zooplankton populations, such as copepods, which are essential for the diet of many larger marine organisms.

Nutrient Availability and Global Change

The relationship between ocean warming and nutrient availability is complex. Warming of the ocean's surface is expected to intensify stratification, creating sharper and more pronounced thermoclines—the transition zones between the warmer mixed layer and the colder deep ocean. This increased stratification, which will last longer throughout the year, acts as a barrier that prevents the upward mixing of cold, nutrient-rich waters into the sunlit mixed layer. Typically, this barrier forms during summertime and breaks down in autumn (see Fig. 32.1). However, with warmer conditions, it may form earlier and break down later in the season, disrupting the nutrient supply to phytoplankton. Since phytoplankton are the primary producers in marine ecosystems, their decline will have cascading effects throughout the entire food web.

Global change may also affect nutrient dynamics in temperate regions in other ways. Increased precipitation and runoff can lead to higher nutrient input from terrestrial sources, which can drive eutrophication. This process often results in harmful algal blooms, which can deplete oxygen levels in the water and produce toxins harmful to marine life.

In the Baltic Sea, for example, increased agricultural runoff has led to high nutrient loads, resulting in frequent and intense blooms of cyanobacteria, such as *Nodularia spumigena*. These blooms produce toxins that can harm fish and other marine organisms, disrupt ecosystem function, and pose risks to human health. The proliferation of such blooms is exacerbated by rising sea temperatures, which favor cyanobacteria over other phytoplankton species.

Part V

Future Directions

33

Plankton and Human Society

If you want happiness for a lifetime, help the next generation
—Chinese saying

At this point in the book, I am confident I convinced you that plankton, despite their microscopic size, play a pivotal role in marine ecosystems and the global environment. However, their importance is often overlooked by the general public. Increasing public awareness and understanding of plankton's crucial role is essential for fostering conservation efforts and ensuring global sustainability. This chapter explores strategies to raise public awareness about plankton and examines the economic consequences of plankton community collapses, all within the framework of plankton's role in global sustainability.

Public Awareness Strategies

Raising public awareness about plankton involves a multifaceted approach that includes education, citizen science, media engagement, and partnerships with educational institutions.

Education: Integrating plankton (or at least ocean-related subjects) education into school curricula is an effective strategy. By teaching students about the role of plankton in marine ecosystems and the global carbon cycle, educators can foster an early appreciation for these vital organisms. Interactive activities, such as plankton sampling and observation under microscopes, can make the learning experience engaging and memorable. This hands-on approach helps students understand the connection between plankton and broader environmental issues like climate change.

Publications: In my opinion, scientists, the real experts on the field, should devote more time to publishing outreach research on plankton and global change, targeting both adults (Fig. 33.1) and children (Fig. 33.2). Increasing the number of accessible publications can deepen our understanding of how these crucial organisms respond to environmental stressors, which is vital for predicting the future health of marine ecosystems and informing conservation efforts. By engaging the public through education and outreach, scientists can raise awareness about the significance of plankton in global biogeochemical cycles and the broader implications of their decline. Educating both adults and children fosters a more informed and supportive society, driving collective efforts to address climate change and protect marine life.

Citizen Science: Citizen science projects also play a crucial role in raising awareness and engaging the public in scientific research. These initiatives not only provide valuable data for researchers but also educate participants about the importance of plankton and the health of marine ecosystems. Engaging citizens in data collection and analysis fosters a sense of stewardship and a personal connection to environmental conservation.

Media Engagement: Whether you love it or hate it, it is undeniable that we live in a world dominated by social media. Media campaigns and public outreach events can further amplify awareness efforts. Documentaries and particularly social media campaigns (Fig. 33.3) and dedicated channels can effectively highlight the significance of plankton and the threats they face due to climate change and pollution. Collaborations with aquariums, museums, and environmental

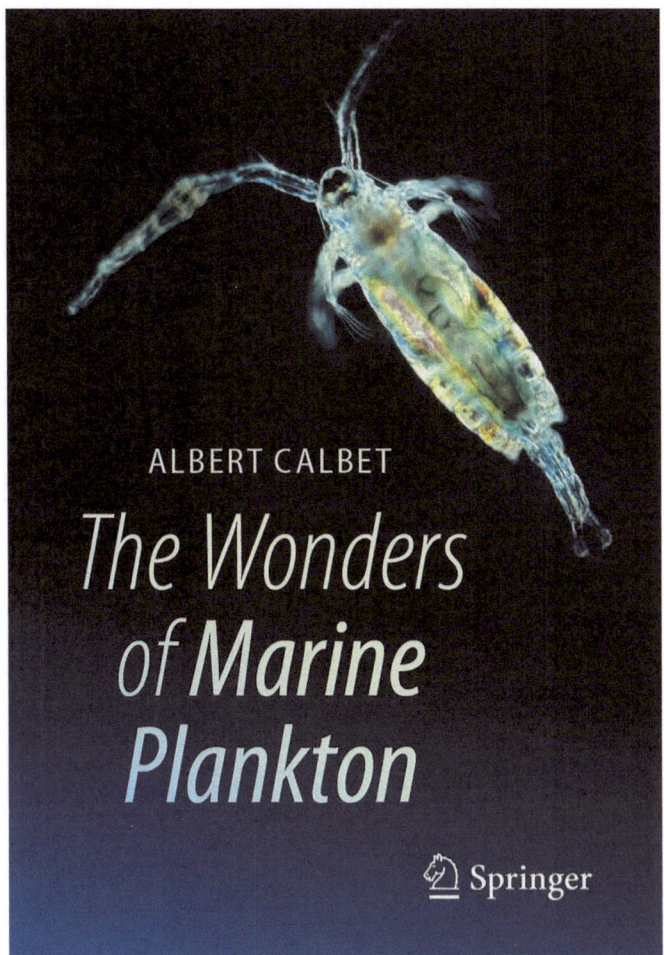

Fig. 33.1 My previous book, *The Wonders of Marine Plankton* (ISBN-13: 978-3031507656), was aimed at the general public and provided an introduction to the ecology, roles, and curiosities of plankton. Therefore, *The Wonders of Marine Plankton* serves as a complementary reading to this book

organizations can provide platforms for public lectures, exhibitions, and interactive displays that educate the public about plankton. By leveraging various media and outreach channels, these efforts can reach a broader audience and inspire action toward marine conservation.

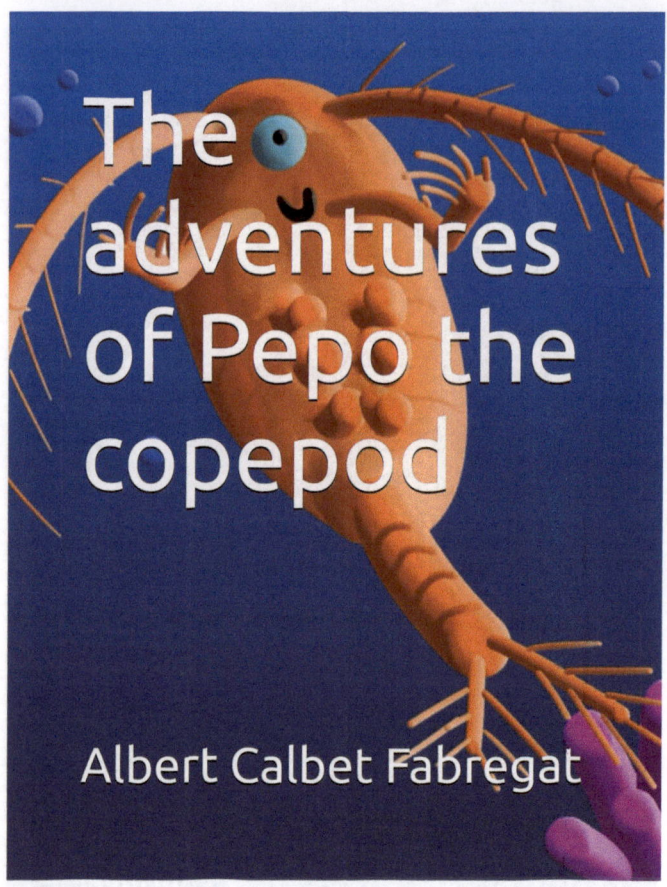

Fig. 33.2 My children's book, *The Adventures of Pepo the Copepod* (ISBN-13: 979-8326818256), is one of the very few of its kind. We often forget that children should be our primary target for raising awareness about the fragility of the marine ecosystem. This book offers two levels of reading: one for children and another for adults, providing more detailed information to help answer children's questions and offering a basic understanding. © Albert Calbet

Economic Consequences of Plankton Community Collapses

The economic impact of plankton community collapses can be devastating, affecting fisheries, tourism, and global food security. Plankton form

Fig. 33.3 Our laboratory's Facebook page is one of the many social media platforms we use to raise awareness about plankton and inform the general public about our recent investigations

the base of marine food webs, supporting a vast array of marine life, including many commercially important fish species. A decline in plankton populations can lead to reduced fish stocks, impacting commercial and recreational fisheries, which are vital sources of income and employment for coastal communities worldwide.

Impact on Fisheries: For instance, as I already mentioned (several times) in previous chapters, the decline of the copepod *Calanus finmarchicus* in the North Atlantic due to rising sea temperatures has had significant repercussions on fish populations such as cod and haddock. These fish rely on *C. finmarchicus* during their larval stages for nutrition. As the copepod populations diminish, the survival rates of fish larvae decrease, leading to lower recruitment and smaller fish stocks. This decline has direct economic consequences for the fishing industry, reducing catches and affecting the livelihoods of fishermen and related industries. Likewise, changes in plankton composition are affecting the fisheries of sardine and anchovy in Mediterranean waters.

Impact on Tourism: The tourism industry can also suffer from plankton community collapses. Marine ecosystems with vibrant plankton populations support diverse and healthy marine life, attracting tourists for activities such as diving, snorkeling, and whale watching. Declines in plankton

Fig. 33.4 Underwater image of coral reef, showing the diversity of forms and colors. Australia, east coast. © Albert Calbet

can lead to less productive and less biodiverse marine environments, making them less attractive to tourists. For example, coral reefs (Fig. 33.4), which are highly dependent on healthy plankton populations for their survival and growth, are major tourist attractions. The degradation of these ecosystems due to declining plankton can lead to a significant loss in tourism revenue. Another common problem associated with plankton community changes is the proliferation of jellyfish, which directly affects tourism.

Impact on Global Food Security: Moreover, the collapse of plankton communities can disrupt global food security. Many developing countries rely heavily on fish as a primary source of protein. Reductions in fish populations due to declining plankton or the accumulation of toxins in seafood from harmful algal blooms can lead to food shortages and increased food prices. This exacerbates food insecurity and malnutrition in vulnerable regions. Consequently, the economic and social stability of these communities can be significantly affected, leading to broader socio-economic challenges.

Plankton in Global Sustainability

Plankton play a critical role in global sustainability by contributing to oxygen production, carbon sequestration, and nutrient cycling. Their role in the biological carbon pump, where carbon is transferred from the surface ocean to the deep sea, helps mitigate the effects of climate change by reducing atmospheric CO_2 levels. This natural carbon sequestration process is essential for maintaining the balance of global carbon cycles and combating global warming.

The health of plankton populations is also indicative of the overall health of marine ecosystems. Monitoring plankton can provide early warnings of environmental changes and potential ecosystem collapses. Efforts to protect and sustain plankton populations are therefore integral to broader conservation and sustainability initiatives. Policies aimed at reducing greenhouse gas emissions, controlling pollution, and managing fisheries sustainably are crucial for preserving plankton communities and ensuring the resilience of marine ecosystems.

International Cooperation and Policy-Making

International cooperation and policy-making play a significant role in protecting plankton and promoting global sustainability. Agreements such as the Paris Climate Agreement (2015) aim to limit global temperature rise and mitigate the impacts of climate change, indirectly benefiting marine ecosystems and plankton populations. Collaborative research and monitoring efforts can enhance our understanding of plankton dynamics and inform conservation strategies.

By recognizing the critical role of plankton in global sustainability and working to raise public awareness, we can better appreciate the interconnectedness of our environmental challenges and take steps toward preserving the health of our planet. Through education, citizen science, media engagement, and international cooperation, we can ensure that these microscopic marvels continue to support the vast and vibrant life in our ocean.

34

Emerging Trends and Future Research Directions in Plankton Studies

Recent technological advances and emerging research trends are revolutionizing the study of plankton, offering new insights and methods to explore their dynamics and interactions. This chapter delves into the latest trends in plankton research, future directions for this field, and the cutting-edge methods for studying and monitoring plankton, including remote sensing and technological innovations.

Emerging Trends in Plankton Research

The field of plankton research is evolving rapidly, driven by the need to understand their role in a changing environment and the development of new technologies. Key emerging trends include the study of plankton biodiversity (Fig. 34.1), the impact of climate change on plankton dynamics, and the role of plankton in marine ecosystem services.

Biodiversity and Functional Roles: Recent research emphasizes the importance of understanding plankton biodiversity and the functional roles of different species. Advances in molecular techniques, such as DNA barcoding and metagenomics, allow scientists to identify and catalog

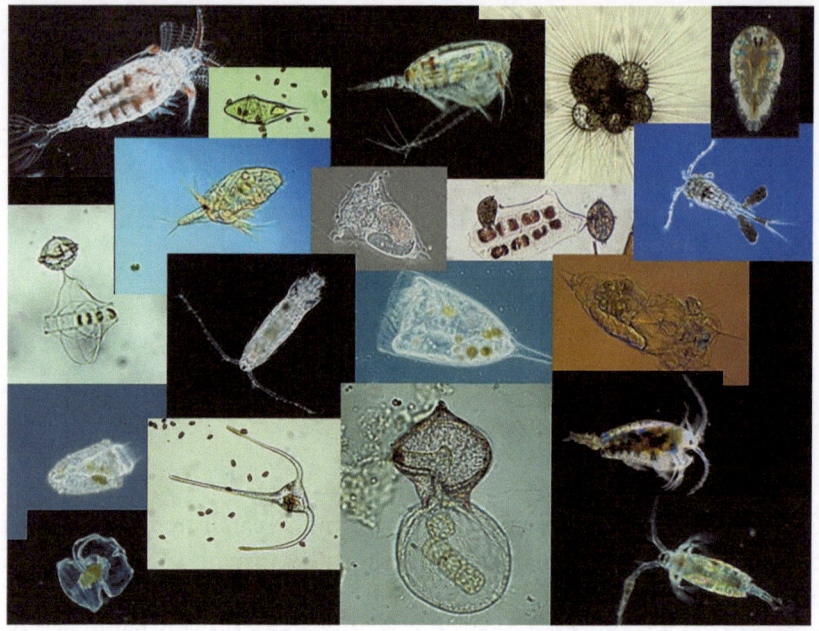

Fig. 34.1 Collage showing the diversity of organisms in zooplankton. © Albert Calbet

plankton species with unprecedented accuracy. These methods have revealed a previously unrecognized diversity of plankton, including many rare and cryptic species that play essential roles in ecosystem functioning.

For example, the Tara Oceans expedition (2009–2013) uncovered an astonishing array of marine life, identifying over 5500 new species of RNA viruses and more than 100,000 new species of single-celled protists, as well as vastly increasing the known diversity of marine bacteria. These discoveries were made possible through advanced techniques like metagenomics and metatranscriptomics, which allowed comprehensive analysis of genetic material from environmental samples, leading to the identification and classification of these microorganisms. Understanding the functional roles of these species is crucial for predicting how plankton communities will respond to environmental changes. Functional traits, such as nutrient uptake efficiency, carbon fixation rates, and predator–prey interactions, determine the ecological roles of different plankton species and their contributions to ecosystem processes.

Impact of Global Change: Global change is profoundly affecting plankton dynamics (Fig. 34.2), with implications for global biogeochemical cycles and marine food webs. Research increasingly focuses on understanding these impacts, including changes in plankton distribution, phenology, and productivity. Warming sea temperatures, ocean acidification, and shifting ocean currents are altering plankton communities, with potential consequences for marine biodiversity and fisheries. Long-term monitoring and predictive modeling are essential for understanding these changes and developing strategies to mitigate their impacts.

Plankton and Marine Ecosystem Services: Plankton provide numerous ecosystem services, including carbon sequestration, oxygen production, and supporting marine food webs. Understanding these services and how they are affected by environmental changes is a growing area of research. For example, studies on the biological carbon pump, driven by plankton are crucial for predicting the ocean's role in mitigating climate change. Research in the North Pacific has demonstrated that diatom blooms

Fig. 34.2 Image of Greenlandic waters during the melting season. The ice melting is an immediate threat for all marine-associated communities of the region. © Albert Calbet

contribute significantly to carbon sequestration, highlighting the importance of protecting these productive plankton communities.

Future Research Directions

The future of plankton research (Fig. 34.3) lies in integrating advanced technologies, interdisciplinary approaches, and global collaboration. Key research directions include new focus on laboratory research, the development of more sophisticated modeling tools, the use of big data and artificial intelligence, and the expansion of global monitoring networks.

New focus on laboratory research: To date, most research on plankton and temperature, among other variables, has been conducted on merely acclimated species. This means that the studies have largely focused on organisms that have been exposed to new conditions for a short period, rather than on those that have undergone long-term adaptation. Present and future research should prioritize understanding the adaptive processes that marine plankton undergo in response to environmental changes. For example, some studies have shown that the metabolic rates of microalgae and copepods reach a balance after proper temperature acclimation, highlighting their capacity for adaptation over extended periods. Such adaptive responses are crucial for predicting how plankton communities will fare in the face of climate change and for developing accurate models of marine ecosystem dynamics.

Moreover, there is a recent tendency to work with trait-based approaches. A trait-based approach in ecology focuses on the functional traits of organisms, such as their morphological, physiological, and behavioral characteristics, rather than on taxonomic identities. These traits determine the organisms' roles in the ecosystem, their interactions with the environment, and their responses to changes. By examining traits like body size, growth rate, feeding strategies, and reproductive methods, researchers can better understand how ecological communities are structured and how they function. This approach allows for more accurate predictions of ecological responses to environmental changes, providing insights into biodiversity, ecosystem dynamics, and resilience. It also facilitates the study of evolutionary processes by linking ecological

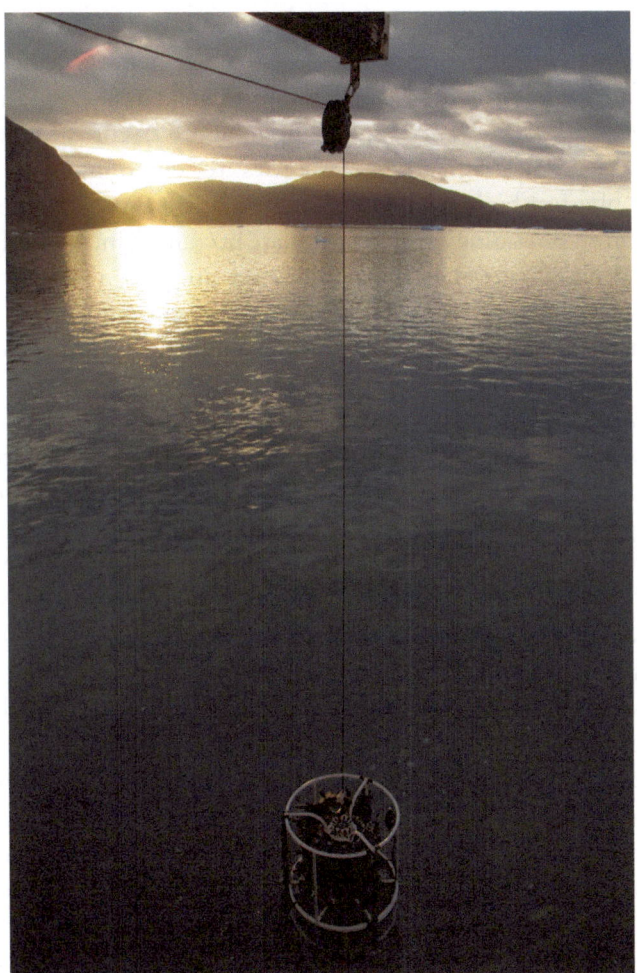

Fig. 34.3 Although not a new tool, the CTD (Conductivity, Temperature, Depth) keeps being one of the most useful tools in oceanographic cruises. A CTD is an oceanographic instrument used to measure the essential physical properties of seawater. © Albert Calbet

performance to specific traits, enabling a comprehensive understanding of how species adapt over time. By linking ecological and evolutionary dynamics, this method provides a comprehensive framework for studying biodiversity and ecosystem function, ultimately aiding in the development of more effective conservation and management strategies.

Modeling and Predictive Tools: Despite their present limitations, discussed in Chap. 28, advanced modeling tools are essential for predicting plankton dynamics and their responses to environmental changes. Coupled physical-biological models that integrate oceanographic data with plankton physiology and behavior are becoming increasingly sophisticated. Although yet they are working in progress, these models can simulate complex interactions within plankton communities and predict how changes in temperature, nutrient availability, and other factors will impact plankton populations.

Big Data and Artificial Intelligence: The integration of big data and artificial intelligence (AI) is transforming plankton research. High-throughput sequencing, automated imaging systems, and remote sensing technologies generate vast amounts of data on plankton communities. AI and machine learning algorithms can analyze these data sets to identify patterns, predict trends, and uncover new insights. For example, automated plankton imaging systems equipped with AI algorithms could rapidly identify and classify plankton species in real-time. This capability enhances the efficiency of monitoring programs and provides high-resolution data on plankton diversity and abundance.

Global Monitoring Networks: Global collaboration and the expansion of monitoring networks are crucial for studying plankton on a large scale. Initiatives such as the Continuous Plankton Recorder (CPR) survey and the Global Alliance of Continuous Plankton Recorder Surveys provide valuable long-term data on plankton trends and distributions. These networks facilitate international collaboration and data sharing, enabling scientists to track changes in plankton communities across different regions and time scales.

The CPR survey, for example, has been instrumental in documenting long-term changes in plankton communities in the North Atlantic. Data from the CPR have shown shifts in plankton species composition and distribution due to climate change, providing critical information for understanding the impacts of environmental changes on marine ecosystems.

Methods for Studying and Monitoring Plankton

Advances in technology are revolutionizing the methods used to study and monitor plankton. Remote sensing, autonomous platforms, and molecular techniques are among the cutting-edge tools that are enhancing our understanding of plankton dynamics.

Remote Sensing: Remote sensing technologies, including satellite observations and oceanographic sensors, provide large-scale and high-resolution data on plankton distributions and environmental conditions. Satellites equipped with ocean color sensors can detect chlorophyll concentrations (a proxy for phytoplankton biomass), among other variables. These data allow scientists to monitor phytoplankton blooms, track changes in productivity, and assess the health of marine ecosystems.

For instance, satellite remote sensing has been used to monitor harmful algal blooms in the Gulf of Mexico. By tracking changes in chlorophyll concentrations, scientists can predict the occurrence of harmful algal blooms and provide early warnings to mitigate their impacts on fisheries and public health.

Autonomous Platforms: Autonomous platforms, such as underwater gliders, drones, and floats, are transforming plankton monitoring by providing continuous and real-time data collection. These platforms can be equipped with various sensors to measure physical, chemical, and biological parameters, offering insights into the complex interactions within marine ecosystems.

For example, underwater gliders deployed in the Southern Ocean have collected data on phytoplankton dynamics, nutrient distributions, and oceanographic conditions. These data have improved our understanding of the drivers of phytoplankton blooms and their role in carbon sequestration.

Molecular Techniques: Molecular techniques, including DNA barcoding, sequencing (Fig. 34.4), metagenomics, and environmental DNA (eDNA) analysis, are revolutionizing the study of plankton biodiversity and community structure. These techniques allow for the identification

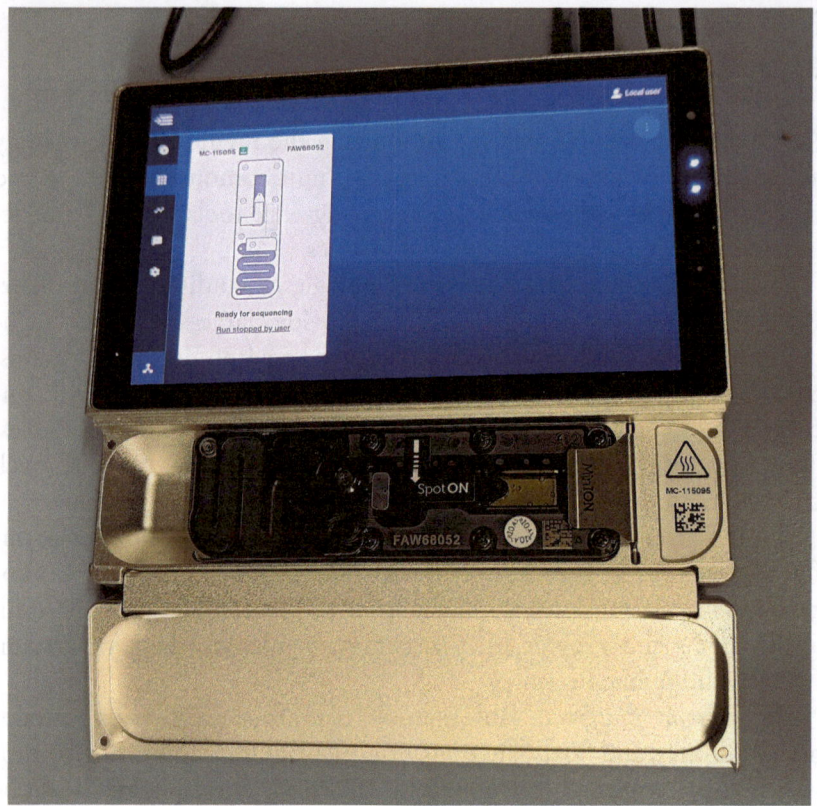

Fig. 34.4 Portable mini-DNA sequencer MinION. This portable device was unimaginable a few years ago

and quantification of plankton species with high precision and can reveal the hidden diversity within plankton communities.

Environmental DNA (eDNA) analysis, for instance, involves sampling water to detect genetic material from plankton. This method can provide comprehensive snapshots of plankton diversity and has been used to monitor plankton communities around the World. The eDNA analysis revealed seasonal changes in plankton species composition and the presence of rare and previously undetected species, enhancing our understanding of marine biodiversity.

The Importance of Classic Methods

While embracing these new technologies, it is crucial not to abandon classic methods, which are essential for a comprehensive understanding of ecosystems. Traditional techniques, such as taxonomy, basic experimentation (Fig. 34.5), and long-term series, provide historical perspectives and baseline data critical for interpreting modern findings and assessing long-term trends. These methods, although sometimes considered slow and labor-intensive, remain invaluable for the continuity and depth of our knowledge about plankton and their ecosystems.

Fig. 34.5 The author of the book in the laboratory, at the Marine Sciences Institute, CSIC, collecting copepods under the stereomicroscope for a routine (yet fascinating) experiment

Hence, my advice for future generations of marine scientists is this: while new technologies may be tempting, never lose your connection with the ocean and its creatures. Strive to know them, understand them, and recognize their limits. This fundamental knowledge is crucial to avoid severe mistakes. Moreover, I have yet to meet anyone who is not captivated by the sight of a live plankton sample under the microscope. The intricate beauty of these tiny organisms can inspire a deep appreciation for the marine world and underscore the importance of its preservation. So, make the effort to engage with and learn about the ocean; it is a source of endless wonder and essential knowledge.

Glossary

Alga   Eukaryotic organism capable of performing oxygenic photosynthesis.

Algal bloom   A rapid and extensive growth of algae in aquatic environments.

Anthropogenic noise   Human-made sounds that disrupt natural soundscapes, impacting marine organisms.

Appendicularia   A small tunicate featuring a tail with a notochord used for propulsion and creating feeding currents.

Artificial Upwelling   Pumping of nutrient-rich deep water to the surface to stimulate plankton growth.

Autotroph   An organism that synthesizes complex organic compounds from carbon dioxide.

Bacterioplankton Bacteria that live in water columns and are crucial for decomposing organic matter.

Biogeochemical Cycles   Natural pathways by which essential elements of living matter are circulated.

Biological Pump The process by which organic carbon is transported from the ocean's surface to the deep sea.

Bioluminescence The production and emission of light by living organisms.

225
A. Calbet, *Plankton in a Changing World*,
https://doi.org/10.1007/978-3-031-76121-8

Bioaccumulation The buildup of pollutants in an organism over time.

Biomagnification The increase in concentration of a pollutant as it moves up the food web.

Calcifying plankton   Plankton that form calcium carbonate shells.

Calanus finmarchicus   A species of copepod found in the North Atlantic and crucial for the sustainment of many fish species larvae, such as cod.

Carbon Cycle   The process by which carbon is exchanged between the earth's ocean, atmosphere, ecosystem, and geosphere.

Citizen science   Scientific research conducted with the participation of non-professional scientists.

Ciliate   A unicellular organism distinguished by the presence of cilia.

Cladoceran   A diminutive freshwater or marine crustacean.

Climate Change Changes in the Earth's climate system, including global warming and other climate patterns.

Climate regulation   The process by which natural systems help control Earth's climate.

CO_2 Abbreviation for carbon dioxide, a significant contributor to global warming.

Competitive Exclusion Principle   The ecological principle stating that two species competing for the same resources cannot coexist at constant population values.

Cyanobacteria   A group of bacteria capable of photosynthesis.

Diatom   A unicellular algae species possessing a siliceous skeleton.

Diel Vertical Migration   Daily movement of plankton from deep waters to the surface at night to feed and descending during the day to avoid predators.

Dinoflagellate   A unicellular organism equipped with two flagella.

Doliolid   A free-living tunicate belonging to the gelatinous plankton.

DNA barcoding A molecular technique for identifying species using a short genetic marker.

Ecosystem services   Benefits provided by ecosystems to humans.

Endocrine-disrupting chemicals (EDCs)   Chemicals that interfere with hormonal systems.

Endospores   Highly resistant dormant stages formed by some bacteria.

Environmental DNA (eDNA)   Genetic material obtained directly from environmental samples.

Epifluorescence microscopy   A microscopy technique that employs light at varying wavelengths to induce fluorescence in the sample.

Ephippium   A protective case for resting eggs produced by some cladocerans.

Eukaryotic   An organism with cells containing a membrane-bound nucleus.

Eutrophication   Excessive nutrient enrichment in water bodies.

Flagellate A unicellular organism propelled by one or more flagella.

Food web An intricate network of feeding interactions defining an ecosystem.

Foraminifera   A group of plankton with calcium carbonate shells.

Geologic time scales   A chronological framework encompassing Earth's history.

Global Change   Large-scale transformations affecting the Earth's systems.

Global warming   The long-term increase in Earth's average surface temperature.

Harmful Algal Blooms (HABs)   Rapid growth of algae that produce toxins harmful to marine life and humans.

Heatwaves   Prolonged periods of abnormally high sea temperatures.

Heterotroph   An organism reliant on consuming organic matter for survival.

Holoplankton   Organisms that complete their entire lifecycle within the plankton community.

Hypoxia   A condition where oxygen levels in water are too low to support most marine life.

Ichthyoplankton The eggs and larvae of fish.

Iron Fertilization The deliberate addition of iron to ocean waters to stimulate phytoplankton growth.

Krill   Euphausiids, small shrimp-like crustaceans.

Marine heatwaves   Periods of extremely high ocean temperatures.

Marine snow   Aggregates of particles occurring in the deep ocean.

Meroplankton   Organisms that spend part of their lifecycle as plankton.

Metagenomics   The study of genetic material recovered directly from environmental samples.

Microplastics   Tiny plastic particles less than 5 millimeters in diameter.

Microzooplankton   Tiny zooplankton, usually protozoans and small metazoans.

Mixoplankton   Plankton that exhibit both plant-like and animal-like behaviors.

Mixotroph   An organism combining autotrophic and heterotrophic feeding strategies.

Mycosporine-like amino acids (MAAs)   Compounds produced by some marine organisms to protect against UV radiation.

Nauplius The larval stage of a crustacean.

North Atlantic Oscillation (NAO) A climate phenomenon influencing weather patterns and ocean circulation in the North Atlantic.

Nutrient Cycling The movement and exchange of organic and inorganic matter.

Nutrient-Phytoplankton-Zooplankton (NPZ) models   Models representing interactions between nutrients, phytoplankton, and zooplankton.

Ocean acidification   The reduction in the pH of the ocean due to increased CO_2 absorption.

Photic zone   The water column layer where sunlight can penetrate.

Phenology   The study of the timing of seasonal biological events.

Photosynthesis The process by which plants use sunlight, water, and carbon dioxide to produce oxygen and energy.

Phytoplankton   Photosynthetic plankton that form the base of the marine food web.

Plankton A collection of mainly small organisms that drift in the water.

Plankton Functional Type (PFT) models   Models categorizing plankton into functional groups.

Primary Consumers   Organisms that feed on primary producers.

Primary Production   The creation of new organic matter by phytoplankton.

Prokaryote   A type of cellular organism that lacks a distinct membrane-bound nucleus.

Prophage A form of a virus integrated into the host's genome.

Protist   A eukaryotic unicellular organism of either animal or plant origin.

Protozoan A eukaryotic unicellular organism of animal or heterotrophic nature.

Redfield Ratio   The typical atomic ratio of carbon, nitrogen, and phosphorus found in marine phytoplankton.

Red tide The accumulation of unicellular algae, often toxic.

Remote sensing The use of satellite or aerial imagery to collect data.

Salp   A tunicate belonging to the gelatinous plankton.

Stratospheric ozone layer A layer in the Earth's stratosphere that absorbs most of the Sun's harmful ultraviolet radiation.

Suspension feeder   An organism that consumes suspended particles.

Symbiotic Relationships   Interactions between different species that live in close physical proximity.

Temperature-Size Rule (TSR)   A rule stating that ectotherms tend to grow to smaller sizes in warmer environments.

Thermocline   A water column layer that demarcates zones with different temperatures.

Trait-Based Approach   An ecological research method focusing on organisms' functional traits—such as morphological, physiological, and behavioral characteristics—rather than their taxonomic identities, to better predict ecological responses and understand biodiversity and ecosystem dynamics.

Tunicate An animal subphylum encompassing ascidians, salps, doliolids, and appendicularians.

Unicellular   Comprising a single cell.

Viral Shunt   The process by which viruses infect and lyse microbial cells.

Virioplankton   Viruses that inhabit marine environments.

Zooplankton   The animal constituents of the plankton community.

Supplementary Bibliography[1]

General Outreach and Reference Books

Calbet A (2024) The wonders of marine plankton. Springer Nature

Roberts C (2012) The ocean of life: the fate of man and the sea. Editorial. Penguin Random House

Castellani C, Edwards M (2017) Marine plankton. A practical guide to ecology, methodology, and taxonomy. Oxford University Press, Oxford

Miller CB, Wheeler PA (2012) Biological oceanography, 2nd edn. Wiley-Blackwell, New Jersey, p 464

Sardet C (ed) (2015) Plankton: wonders of the drifting world. The University of Chicago Press, Chicago, p 222

---

[1] This book is designed to inspire curiosity and educate readers about the importance, complexity, and fragility of plankton and the threats of global change. It is not intended as a reference or textbook, so I have chosen not to include scientific references for each of the arguments presented in the text. For those who wish to dive deeper into the science of plankton, I have included a short list of scholarly articles and accessible books that offer extensive insights into this fascinating subject.

© The Author(s), under exclusive license to Springer Nature Switzerland AG 2024
A. Calbet, *Plankton in a Changing World*,
https://doi.org/10.1007/978-3-031-76121-8

More Specialized Articles and Books: Chapter 1: An Introduction to the World of Marine Plankton

Pomeroy LR (1974) The ocean's food web, a changing paradigm. Bioscience 24(9):499–504

Azam F, Fenchel T, Field JG, Gray JS, Meyer-Rei LA, Thingstad F (1983) The ecological role of water-column microbes in the sea. Mar Ecol Prog Ser 10:257–263

More Specialized Articles and Books: Chapter 2: Plankton Size Classification

Sieburth JMN, Smetacek V, Lenz J (1978) Pelagic ecosystem structure: heterotrophic compartments of the plankton and their relationship to plankton size fractions. Limnol Oceanogr 23(6):1256–1263

More Specialized Articles and Books: Chapter 3: From the Bottom Up: Phytoplankton's Major Groups

Falkowski PG, Raven JA (2007) Aquatic photosynthesis. Princeton University Press

Field CB, Behrenfeld MJ, Randerson JT, Falkowski P (1998a) Primary production of the biosphere: integrating terrestrial and oceanic components. Science 281(5374):237–240

More Specialized Articles and Books: Chapter 4: Ocean's Hidden Gardeners: The World of Underwater Micro-Grazers

Calbet A (2008) The trophic roles of microzooplankton in marine systems. ICES J Mar Sci 65:325–331

Mauchline J (1998) The biology of calanoid copepods. Academic Press

Sanders RW (1991) Mixotrophic protists in marine and freshwater ecosystems. J Protozool 38(1):76–81

Stoecker DK, Michaels AE (1991) Mixotrophy in marine planktonic ciliates: physiological and ecological aspects of plastid-retention. J Eukaryot Microbiol 38(4):251–258

More Specialized Articles and Books: Chapter 5: Bacteria and Archaea: The Crucial Roles of Prokaryotes in Ocean Environments

DeLong EF, Karl DM (2005) Genomic perspectives in microbial oceanography. Nature 437(7057):336–342

Falkowski PG, Fenchel T, Delong EF (2008) The microbial engines that drive Earth's biogeochemical cycles. Science 320(5879):1034–1039

Whitman WB, Coleman DC, Wiebe WJ (1998) Prokaryotes: the unseen majority. Proc Natl Acad Sci 95(12):6578–6583

More Specialized Articles and Books: Chapter 6: Tiny Terrors of the Ocean: Planktonic Viruses and Parasites

Fuhrman JA (1999) Marine viruses and their biogeochemical and ecological effects. Nature 399(6736):541–548

Suttle CA (2005) Viruses in the sea. Nature 437(7057):356–361

Weinbauer MG (2004) Ecology of prokaryotic viruses. FEMS Microbiol Rev 28(2):127–181

More Specialized Articles and Books: Chapter 7: The Hidden World of Marine Fungi

Baltar F (ed) (2022) Marine fungus. MDPI AG

Richards TA, Jones MDM (2012) Marine fungi: their ecology and molecular diversity. Annu Rev Mar Sci 4:495–522

More Specialized Articles and Books: Chapter 8: The Biomass Distribution in Earth's Ecosystems

Bar-On YM, Phillips R, Milo R (2018) The biomass distribution on earth. Proc Natl Acad Sci 115(25):6506–6511
Field CB, Behrenfeld MJ, Randerson JT, Falkowski P (1998b) Primary production of the biosphere: integrating terrestrial and oceanic components. Science 281(5374):237–240
Longhurst AR (1998a) Ecological geography of the sea. Academic Press

More Specialized Articles and Books: Chapter 9: A Patchy Ocean: From Microscopic Meals to Macroscale Aggregations

Kiørboe T (2008) A mechanistic approach to plankton ecology. Princeton University Press
Martin AP (2003) Phytoplankton patchiness: the role of lateral stirring and mixing. Prog Oceanogr 57(2):125–174

More Specialized Articles and Books: Chapter 10: Plankton Across Ecosystems

Behrenfeld MJ, Boss ES (2014a) Resurrecting the ecological underpinnings of ocean plankton blooms. Annu Rev Mar Sci 6:167–194
Falkowski PG, Oliver MJ (2007) Mix and match: how climate selects phytoplankton. Nat Rev Microbiol 5(10):813–819
Longhurst AR (1998b) Ecological geography of the sea. Academic Press

More Specialized Articles and Books: Chapter 11: Adrift but Not Lost: Dispersal and Colonization Strategies of Plankton

Cowen RK, Gawarkiewicz G, Pineda J, Thorrold SR, Werner FE (2007) Population connectivity in marine systems: an overview. Oceanography 20(3):14–21

Haury LR, McGowan JA, Wiebe PH (1978) Patterns and processes in the time-space scales of plankton distributions. In: Patterns in plankton ecology, pp 277–327

More Specialized Articles and Books: Chapter 12: Sailing Through Chaos: Turbulence's Impacts on Plankton

Kiørboe T (1997) Small-scale turbulence, marine snow formation, and planktivorous feeding. Sci Mar 61(S1):141–158

Rothschild BJ, Osborn TR (1988) Small-scale turbulence and plankton contact rates. J Plankton Res 10(3):465–474

More Specialized Articles and Books: Chapter 13: Plankton Sentinels: Resting Cysts and Dormancy

Anderson DM (1989) Toxic algal blooms and red tides: a global perspective. In: Red tides: biology, environmental science, and toxicology, pp 11–16

Fryxell GA (1983) Survival strategies of the algae. Cambridge University Press

More Specialized Articles and Books: Chapter 14: The Paradox of Plankton: The Kill the Winner Hypotheses

Thingstad TF (2000) Elements of a theory for the mechanisms controlling abundance, diversity, and biogeochemical role of lytic bacterial viruses in aquatic systems. Limnol Oceanogr 45(6):1320–1328

Winter C, Bouvier T, Weinbauer MG, Thingstad TF (2010) Trade-offs between competition and defense specialists among unicellular planktonic organisms: the 'killing the winner' hypothesis revisited. Microbiol Mol Biol Rev 74(1):42–57

More Specialized Articles and Books: Chapter 15: Understanding Global Change, Global Warming, and Climate Change

Field CB, Barros VR, Dokken DJ, Mach KJ, Mastrandrea MD, Bilir TE et al (2014) Climate change 2014: impacts, adaptation, and vulnerability. Part A: global and sectoral aspects. Contribution of Working Group II to the Fifth Assessment Report of the Intergovernmental Panel on Climate Change. Cambridge University Press

IPCC (2023) Climate change 2023: synthesis report. Contribution of Working Groups I, II and III to the Sixth Assessment Report of the Intergovernmental Panel on Climate Change [Core Writing Team, H. Lee and J. Romero (eds.)]. IPCC, Geneva, pp 35–115. https://doi.org/10.59327/IPCC/AR6-9789291691647

Pachauri RK, Meyer L, Plattner GK (2014) Climate change 2014 synthesis report: summary for policymakers. IPCC

More Specialized Articles and Books: Chapter 16: The Impacts of Global Warming on Marine Plankton

Boyce DG, Lewis MR, Worm B (2010) Global phytoplankton decline over the past century. Nature 466(7306):591–596

Hays GC, Richardson AJ, Robinson C (2005) Climate change and marine plankton. Trends Ecol Evol 20:337–344

Hoegh-Guldberg O, Bruno JF (2010) The impact of climate change on the world's marine ecosystems. Science 328(5985):1523–1528

Poloczanska ES, Brown CJ, Sydeman WJ, Kiessling W, Schoeman DS, Moore PJ et al (2013) Global imprint of climate change on marine life. Nat Clim Chang 3(10):919–925

More Specialized Articles and Books: Chapter 17: The Consequences of Smaller Plankton in a Warmer Ocean

Forster J, Hirst AG, Atkinson D (2012) Warming-induced reductions in body size are greater in aquatic than terrestrial species. Proc Natl Acad Sci 109(47):19310–19314. https://doi.org/10.1073/pnas.1210460109

Gardner JL, Peters A, Kearney MR, Joseph L, Heinsohn R (2011) Declining body size: a third universal response to warming? Trends Ecol Evol 26(6):285–291. https://doi.org/10.1016/j.tree.2011.03.005

More Specialized Articles and Books: Chapter 18: The Ocean's Biological Pump: A Crucial Ally Against Global Warming

Buesseler KO, Boyd PW (2009) Shedding light on processes that control particle export and flux attenuation in the twilight zone of the open ocean. Limnol Oceanogr 54(4):1210–1232

Passow U, Carlson CA (2012) The biological pump in a high CO_2 world. Mar Ecol Prog Ser 470:249–271

Volk T, Hoffert MI (1985) Ocean carbon pumps: analysis of relative strengths and efficiencies in ocean-driven atmospheric CO_2 changes. Geophys Monogr Ser 32:99–110

More Specialized Articles and Books: Chapter 19: Blue Carbon Revolution: Exploiting Plankton to Combat CO₂ Emissions

Howard J, Hoyt S, Isensee K, Telszewski M, Pidgeon E (2014) Coastal blue carbon: methods for assessing carbon stocks and emissions factors in mangroves, tidal salt marshes, and seagrass meadows. Conservation International

Macreadie PI, Anton A, Raven JA, Beaumont N, Connolly RM, Friess DA et al (2019) The future of blue carbon science. Nat Commun 10(1):3998

More Specialized Articles and Books: Chapter 20: The Impact of Large-Scale Climatological Events on Plankton Populations

Beaugrand G, Reid PC, Ibañez F, Lindley JA, Edwards M (2002) Reorganization of North Atlantic marine copepod biodiversity and climate. Science 296(5573):1692–1694

Chavez FP, Strutton PG, Friederich GE, Feely RA, Feldman GC, Foley DC, McPhaden MJ (1999) Biological and chemical response of the equatorial pacific ocean to the 1997-98 El Niño. Science 286(5447):2126–2131

More Specialized Articles and Books: Chapter 21: Ocean Acidification and Plankton: The Unnoticed Crisis

Doney SC, Fabry VJ, Feely RA, Kleypas JA (2009) Ocean acidification: the other CO₂ problem. Annu Rev Mar Sci 1:169–192

Riebesell U, Schulz KG, Bellerby RG, Botros M, Fritsche P, Meyerhöfer M et al (2007) Enhanced biological carbon consumption in a high CO₂ ocean. Nature 450(7169):545–548

More Specialized Articles and Books: Chapter 22: Nutrient Availability and Its Consequences for Plankton in a Changing World

Cloern JE (2001) Our evolving conceptual model of the coastal eutrophication problem. Mar Ecol Prog Ser 210:223–253

Moore CM, Mills MM, Arrigo KR, Berman-Frank I, Bopp L, Boyd PW et al (2013) Processes and patterns of oceanic nutrient limitation. Nat Geosci 6(9):701–710

Smith VH (2003) Eutrophication of freshwater and coastal marine ecosystems: a global problem. Environ Sci Pollut Res 10(2):126–139

More Specialized Articles and Books: Chapter 23: Global Change and Salinity: Implications for Plankton

Hall CAM, Lewandowska AM (2022) Zooplankton dominance shift in response to climate-driven salinity change: a mesocosm study. Front Mar Sci 9:861297. https://doi.org/10.3389/fmars.2022.861297

Hopwood MJ, Carroll D, Dunse T, Hodson A, Holding JM, Iriarte JL, Ribeiro S, Achterberg EP, Cantoni C, Carlson DF, Chierici M, Clarke JS, Cozzi S, Fransson A, Juul-Pedersen T, Winding MHS, Meire L (2020) How does glacier discharge affect marine biogeochemistry and primary production in the Arctic? Cryosphere 14:1347–1383

More Specialized Articles and Books: Chapter 24: The Underappreciated Threat of Ultraviolet Radiation to Marine Plankton

Häder DP, Kumar HD, Smith RC, Worrest RC (2007) Effects of solar UV radiation on aquatic ecosystems and interactions with climate change. Photochem Photobiol Sci 6(3):267–285

Villafañe VE, Helbling EW (2005) Short-term acclimation responses of the marine diatom Thalassiosira weissflogii to UV radiation. Mar Ecol Prog Ser 290:35–44

More Specialized Articles and Books: Chapter 25: The Silent Killers: Classic and Emerging Pollutants and Marine Plankton

Almeda R, Wambaugh Z, Buskey EJ (2013) Effects of crude oil exposure on swimming performance and respiration of the copepod *Acartia tonsa*. Environ Pollut 181:271–278

Wright SL, Thompson RC, Galloway TS (2013) The physical impacts of microplastics on marine organisms: a review. Environ Pollut 178:483–492

Xu J, Rodríguez-Torres R, Rist S, Nielsen TG, Hartmann NB, Brun P, Li D, Almeda R (2022) Unpalatable plastic: efficient taste discrimination of microplastics in planktonic copepods. Environ Sci Technol 56(10):6455–6465

More Specialized Articles and Books: Chapter 26: The Unseen Threats: Effects of Sound and Light Pollution on Plankton

Peng C, Zhao X, Liu G (2015) Noise in the sea and its impacts on marine organisms. Int J Environ Res Public Health 12(10):12304–12323

Marangoni LFB, Davies T, Smyth T, Rodríguez A, Hamann M, Duarte C, Pendoley K, Berge J, Maggi E, Levy O (2022) Impacts of artificial light at night in marine ecosystems—a review. Glob Chang Biol 28:5346–5367

More Specialized Articles and Books: Chapter 27: Plankton and Global Fisheries

Beaugrand G, Brander KM, Lindley JA, Souissi S, Reid PC (2003) Plankton effect on cod recruitment in the North Sea. Nature 426(6967):661–664

Checkley DM, Barth JA (2009) Patterns and processes in the California current system. Prog Oceanogr 83(1-4):49–64

Richardson AJ, Schoeman DS (2004) Climate impact on plankton ecosystems in the Northeast Atlantic. Science 305(5690):1603–1606

More Specialized Articles and Books: Chapter 28: Predicting the Future of Our Seas and Ocean Ecosystems: Reality or Chimera?

Boyd PW, Doney SC (2002) Modeling regional responses by marine pelagic ecosystems to global climate change. Geophys Res Lett 29(16):53-1–53-4

Doney SC, Ruckelshaus M, Duffy JE, Barry JP, Chan F, English CA et al (2012) Climate change impacts on marine ecosystems. Annu Rev Mar Sci 4:11–37

Stock CA, Dunne JP, John JG (2014) Global-scale carbon and energy flows through the marine planktonic food web: an analysis with a coupled physical–biological model. Prog Oceanogr 120:1–28

More Specialized Articles and Books: Chapter 29: The Plankton Time Machine: A Journey Through Geological Eras

Falkowski PG, Katz ME, Knoll AH, Quigg A, Raven JA, Schofield O, Taylor FJR (2004) The evolution of modern eukaryotic phytoplankton. Science 305(5682):354–360

Pearson PN, Palmer MR (2000) Atmospheric carbon dioxide concentrations over the past 60 million years. Nature 406(6797):695–699

More Specialized Articles and Books: Chapter 30: The Effects of Global Change on Plankton in Polar Regions

Arrigo KR, van Dijken G, Pabi S (2008) Impact of a shrinking Arctic ice cover on marine primary production. Geophys Res Lett 35(19):L19603

Leu E, Wiktor JM, Søreide JE, Berge J, Falk-Petersen S (2010) Increased irradiance reduces food quality of sea ice algae. Mar Ecol Prog Ser 411:49–60

More Specialized Articles and Books: Chapter 31: The Effects of Global Change on Plankton in Tropical Regions

Behrenfeld MJ, O'Malley RT, Siegel DA, McClain CR, Sarmiento JL, Feldman GC et al (2006) Climate-driven trends in contemporary ocean productivity. Nature 444(7120):752–755

Hoegh-Guldberg O (1999) Climate change, coral bleaching and the future of the world's coral reefs. Mar Freshw Res 50(8):839–866

Hutchins DA, Fu F, Zhang Y, Warner ME, Feng Y, Portune K et al (2007) CO_2 control of *Trichodesmium* N_2 fixation, photosynthesis, growth rates, and elemental ratios: implications for past, present, and future ocean biogeochemistry. Limnol Oceanogr 52(4):1293–1304

More Specialized Articles and Books: Chapter 32: The Effects of Global Change on Plankton in Temperate Regions

Cloern JE, Foster SQ, Kleckner AE (2014) Phytoplankton primary production in the world's estuarine-coastal ecosystems. Biogeosciences 11(9):2477–2501

Moore JK, Doney SC, Glover DM, Fung IY (2002) Iron cycling and nutrient-limitation patterns in surface waters of the World Ocean. Deep-Sea Res II Top Stud Oceanogr 49(1-3):463–507

Sommer U, Lewandowska A (2011) Climate change and the phytoplankton spring bloom: warming and overwintering zooplankton have similar effects on phytoplankton. Glob Chang Biol 17(1):154–162

More Specialized Articles and Books: Chapter 33: Plankton and Human Society

Falkowski PG, Barber RT, Smetacek V (1998) Biogeochemical controls and feedbacks on ocean primary production. Science 281(5374):200–206

Behrenfeld MJ, Boss ES (2014b) Resurrecting the ecological underpinnings of ocean plankton blooms. Annu Rev Mar Sci 6:167–194

Le Quéré C, Harrison SP, Colin Prentice I (2009) Ecosystem dynamics based on plankton functional types for global ocean biogeochemistry models. Glob Chang Biol 11(11):2016–2040

More Specialized Articles and Books: Chapter 34: Emerging Trends and Future Research Directions in Plankton Studies

De Vargas C, Audic S, Henry N, Decelle J, Mahé F, Logares R et al (2015) Eukaryotic plankton diversity in the sunlit ocean. Science 348(6237):1261605

Guidi L, Chaffron S, Bittner L (2016) Plankton networks driving carbon export in the oligotrophic ocean. Nature 532(7600):465–470

Litchman E, Klausmeier CA (2022) Future directions in plankton ecology: understanding trait-based ecology and evolution in plankton. Annu Rev Mar Sci 14:1–21. https://doi.org/10.1146/annurev-marine-020821-100536